Radical
Ecological
Thought
Experiments

Books from Mozart & Reason Wolfe Ltd.

Urania Science Press
RE: viewing thinking & turning
Good Forestry from Good Theories & Good Practices
(O)utopias or (E)utopias, by Alan Wittbecker
Topopoetics, by Alan Wittbecker
Global Emergency Actions, by Alan Wittbecker
Redesigning the Planet: Foundations
Redesigning the Planet: Local Systems
Redesigning the Planet: Regions
Redesigning the Planet: Global Ecological Designs

Calliope Press
Two Diaries, by Marcus Rian
Lucifer Dreaming, by Marcus Rian
Amphibian Dreams, by A. M. Caratheodory
Wild Apples, by A. M. Caratheodory
Light from a Vanished Forest, by A. M. Caratheodory
Cheap Visions, by Violet Reason
Carbon Dreams, by Violet Reason
Tropomorphoses, by Yulalona Lopez
Night Wolves, by Yulalona Lopez
Shadow Masks, Ed. by Crawford Washington

Clio Press
The Thesis, by Marcus Rian
Waiting for Better Times in Bulgaria, by Conor Ciaran
Musings, Ed. by Crawford Washington
Murmur, Ed. by Crawford Washington
Guarded by Trees, by A. M. Caratheodory
Coyote Redux, by Violet Reason & Yulalona Lopez
Coyote Remasked, by Yulalona Lopez
Coyote Renegade, by Yulalona Lopez
Coyote Refocused, by Yulalona Lopez
Poetic Archaeology of the Flesh, by Alan Wittbecker
One Earth Many Worlds, by Alan Wittbecker
Global Government, by Alan Wittbecker
Eutopias: Making Good Places Ecologically & Culturally

Radical
Ecological
Thought
Experiments

Including Mini and Micro Thoughts
On Various Ponderous Concerns
To Support the Irrefutable Conclusion
That the Planet is Suffering
An Emergency *Now!*

Urania Science Press
Mozart & Reason Wolfe, Ltd.
Sarasota Florida 2015

In memoriam: Arne Naess, Paolo Soleri, R. Buckminster Fuller, Garrett Hardin Merv Wilkinson, Paul Shepard, Leopold Kohr, Thomas Berry, Ivan Illich, Emerson Wittbecker

Cover Design: 2015, Rian Garcia Calusa
Graphics and photographs: Alan Wittbecker (unless noted in text)

Published by Urania Science Press
for Mozart & Reason Wolfe, Ltd.
at SynGeo ArchiGraph
Mail: Post Office Box 370, Tallevast, Florida 34270
Email: Design@SynGeo.org

Publisher's Cataloging in Publication Data
Alan Wittbecker 1946—
Radical Ecological Thought Experiments / Alan Wittbecker

Edition Version 1.2

Library of Congress Control Number:

ISBN-13: 978-1500230319
ISBN-10: 1500230316

1. Human Ecology. 2. Thought Experiments. 3. Essays.
I. Title. GF78.W54 2015

Book Design by Rian Garcia Calusa, Sarasota, Florida

Printed in the United States of America

Table of Contents

Dedication

    Michael W. Fox, John B. Cobb Jr, Paolo Soleri
    Arne Naess, Leopold Kohr, Paul Shepard
    Neil Everenden, Henryk Skolimowski, Alan Drengson
    David Klein, R. Buckminster Fuller–for all their good
    ideas.

    Thanks to Marcella, Emerson, and Margaret for reading
    and criticizing some of the experiments. Thanks to TB,
    TJ, CW, and NK for their encouragement.

Author's Note

Most of these experiments and ideas have been suggested by other authors from the 1860s to the present. Many authors have explored these areas in detail, but few have taken the step to assemble all of the pieces into one set of interlinked actions (as in *Global Emergency Actions* or *Redesigning the Planet*). This work is an attempt to distill the experimental parts to highlight key actions to avoid catastrophes.

I regret that most of my work has not as been as concise as the works that inspired me, but I am still learning to focus on one idea, to write simply, and to edit it down. Some of these essays reflects the limitations of some journals and newspaper columns, which had space requirements or 1000-word limits.

The crucial message is that we are facing a number of large impending catastrophes that are converging and will require substantial coordinated emergency actions that we may not be able to manage with our current national and international structures, or with our fractured technohumanistic mindset, or because of the fragile poverties that are sapping our abilities to respond appropriately and quickly. This is one massive global emergency, and if we do not work together, we may not survive as a civilization. We may be reduced to tribal groups for a while, or to a few small enclaves of rich surrounded by and raided by those more desperate. I would much rather participate in a grand experiment with everyone else than just be overwhelmed by the catastrophes we should have seen coming for the past hundred years. Let's start together now and build a system that not only survives, but thrives and continues developing. *Now, it is an emergency!*

Introduction: Radical Ecological Thought Experiments?

Inspired by the writings of Buckminster Fuller, John B. Cobb Jr, Paolo Soleri, and Paul Shephard in 1969, I outlined a project that would expand Fuller's distinction and use of the concept of Eutopias (An alternative to Thomas More's clever pun on Outopias and Eutopias to become Utopias) for the entire planet, by using traditional cultures as the unit of independence, and by allowing Leopold Kohr's breakdown of nations to shape those units.

I tried to answer Cobb's recognition of civilization's critical closeness to catastrophe by using Soleri's efforts to create arcologies, ecological cities, as a solution to the urban spread of a technological species. This was combined with the conservation and restoration of large areas of the planet, which would be left wild, as Paul Shepard recommended. This could be possible if humans expressed a reverence for life as a result of a biospiritual ethic, as Michael W. Fox has urged, and if they adopted a catastrophic psychology to implement it as an immediate series of emergency actions. I then immersed myself in the works of Leopold Kohr, Michael W. Fox, Henryk Skowlimowski, Alan Drengson, and Arne Naess. And, I rediscovered Einstein and Infield, and a framework to present ideas.

One way to visualize these actions is as thought experiments. In their book *The Evolution of Physics*, A. Einstein and L. Infield, suggest that knowledge of the laws of nature can be gained through the contemplation of idealized experiments created by thought, Gedanke-Experiment. For example, to address the equality of inertial and gravitational masses, Einstein imagined an elevator at the top of an incredibly high building, and then imagined what research would be done in this local moving environment. Such experiments might seem "fantastic" he said, but they might help us understand what we want to understand. Although ecology and global actions are orders of magnitude more complex than physical systems, perhaps we could imagine and use such experiments to help us understand what is happening with our complex, living planet, which is composed of many interlocking ecological, physical, and political systems. By asking all kinds of questions and then imagining the answers, given what we know about our history and ourselves, we could discuss things that are often taken for granted, or not even thought about.

What is radical about these thought experiments? Radical in this sense simply means being rooted, returning to grounded ideas and common sense. These experiments recommend revolutionary change. Unfortunately, revolutions have the connotations of violence and overthrow. Revolutions can be as quiet and regular, and unthreatening, as the turnover of an axle on a wagon or car. Thomas Jefferson suggested that little revolutions, every couple of decades in the US, could make the experiment fresh, as well as break up unproductive hierarchies of power. We could start these little revolutions with many small steps, as long as they did not contradict cultural norms. The first revolution has to be to adopt a new attitude. Adopting a catastrophic attitude for the nation, to address current and imminent losses, is less revolutionary and a more appropriate response to large scale catastrophes, such as the national loses of biological and cultural heritages.

Do not worry that change cannot be sudden or that people cannot adjust. Do not believe that plans have to be slow and long-range. They don't. Industrial progress has presented us with sudden change with its accelerated use of oil and cheap labor. Archaic cultures and middle-class communities are replaced or destroyed overnight. Public power or control of production has been reduced or eliminated. Putting off our comfort and habits and postponing a long-term effort will be meaningless if we continue our genocidal military adventures in collapsing ecosystems under a hot atmosphere.

What is ecological about them? They are based on an ecological perspective that is used to inform the operation of thinking, by tracing the connections of living things to reveal a network that is spread around the planet, backwards and forwards in time. Paul Shepard said that through its components and relationships, which are as real as the components, ecology is a way of seeing, a perspective that goes beyond the science to become a 'subversive' vision that is sensible and normative.

In this way, we can see that humanity has grown through this network, touching and experimenting with the connections between plants and animals, adding some and removing many others, replacing many with human beings. Humanity has been experimenting with the environment, and these experiments added up until we were experimenting with the entire planet. We have so many unplanned experiments, it is worrisome. Urbanization is a tremendous unplanned, undesigned, and unmonitored experiment. And, it involves over half the population of the planet. The conversion of wild ecosystems into fields for grazing or growing single crops is an equally tremendous experiment, and it's not remotely scientific. The Internet to connect the majority of computers on the planet is an exciting, uncontrolled experiment, not only in communication, but in virtual worlds, political transparency and identity theft as well. So these small suggested experiments should seem more rational and limited in their approach than what we are doing blindly, now!

Insightful change cannot be worse than blind progress. So, let's take immediate action, using the values and knowledge in traditional cultures, combined with enhanced ecological knowledge and perspectives, and a new enlightened global approach that would minimize conflict and war by returning much power to local levels and allowing an empowered United Nations address all international conflicts and common problems, from national failures to atmospheric heating. The transformation has to be coordinated and complete; it has to be flexible and pragmatic. Because of our global connections it cannot be half done, or planned slowly for later implementation.

Of course, there will be some suffering and some waste. But, it will be less than the current waste and suffering. Think not? Ask the insulted, polluted, poor, dying, starving, and homeless billions. I believe that they would respond well to your interest. The millions of currently rich and powerful are going to be rabid about some changes, especially economic and political ones, but they will still have wealth and power. Getting rid of massive social inequality and fiscal inequity will still leave dramatic economic and social differences, but it will allow new leveling and more balanced opportunities. Laws are made by legislatures elected by groups to represent them. This might make it hard to pass laws that transfer wealth or address global phenomena, since the wealthy can influence elections and laws. This is another reason why the United Nations needs more power, to control the commons and to disarm large nations. People will still make bad decisions and mistakes, but on a smaller scale, they can be corrected faster and more easily. And, if these actions do not work, we can always go back to blind, unconscious growth. That is the beauty of experiments—if they fail, we can learn from them and try another.

Can we end the accidental experiments, though, so we can proceed with considered and cautious, but ecological and revolutionary experiments? We need these to firmly address the long emergencies of cultural and national weakness, and environmental threats. The value of these experiments is that they put us all together on an emergency footing, quite similar to wartime, when people accept the necessity of job change, equal sacrifice, rationing, and comradery from having to work for a common goal. In this case the real goal is to keep the entire planet, with all its varieties of life, including human, healthy and expressive, or in Michael W. Fox's fine words, it is "To restore the song of the earth."

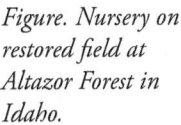

Figure. Nursery on restored field at Altazor Forest in Idaho.

GP Marsh Institute Series 1970—

Recognize Small Nations

The United Nations would formally recognized any former nation or traditional culture, and permit them join the UN with all the rights and advantages of any current nation. This would solve some of the problems facing many nations, especially those who were formed violently by neighbors or from the colonial enterprise over the past 500 years. The new nation could still be a member of a larger association designated Uganda, France, Russia, or China—and those nations could still exist as smaller territorial entities and as large associations.

The current large nations may be unhappy at first with such a situation, due to their perceived loss of weight, importance, and power. But, they might recognize the advantages, especially in terms of lower costs and fewer problems with misunderstandings and violence. The current nations have not created good places; the global market has not been able to create health and equity for populations; even scientists have not been able to create a way of restoring the health of ecosystems or the planet.

What would be the consequences of this experiment? It might detoxify national or ethnic rivalries. As Leopold Kohr argued, when he made this proposal, the former large nations, now associations, might have a stronger presence in the UN because the independent members of their associations would have separate voting rights.

It would be a good for small nations based on traditional cultures to be finally represented in the UN. Racism, sexism and ageism could lose their importance in a smaller, cooperative nation of advanced communication, equality, humane scale, and meaningful preservation.

This would give the small nations a chance to try to build progressive, partially-planned societies. A small nation might be more likely to accept the imperfect nature of humanity and the changing ambiguity of nature. It would be a framework for local self-reliance and global exchange, respectful of cultural and ecological limits and networks.

The UN would become larger, increasing to over a thousand members, perhaps close to three thousand members. It would be a framework where different human experiments were tried. Its variability would insure that we could reject any of the local visions that fail. This change requires a change in awareness so that we can recognize that we are inseparable from each other (and from the environment), sharing a common fate, and being responsive to catastrophes. This experiment should be tried at once, in recognition of the emergencies facing civilization.

Give More Power to the United Nations

More power would make the UN into a true global institution. Let it rewrite its charter to go beyond the original activities of human rights, education and peacekeeping. After all, it represents all people, and should include all living beings. It could:

- *Create new Charter and Constitution.* Create new structures and branches, for instance, Offices of Standards, Budget, Normalization, and Taxation, and Ministries of Biosphere Health and Resource Management. The UN would take ownership of all global commons, such as the Ocean, and global resources, such as oil. It would levy taxes (transitional or permanent) on the use of all cycles and elements.
- *Give the UN power to disperse all armies and disarm them of large weapons*, especially nuclear or biological. Only police powers would be left to nations. Nations have been promising to give up all nuclear weapons. Disarmament could be finished in one week, according to Earl Osborn.
- *Create a security force* to protect nations and the planet, meaning all ecosystems and planetary cycles. This force would be larger than the combined forces of any three large nations. It would also be responsible for removing abandoned ordinance, including land mines.
- *Implement a formal Global Ecological Survey* of all ecological systems and cycles.
- *Create new agencies* for food & shelter, energy, technology & communications, education and research, business & corporations, and other things. All landless corporations would be tied to a nation, with new responsibilities and limits. One agency could incorporate the planet.
- *Create new offices*, including Personnel & staff, Self-assessment & the Future, Ombudsman, and Volunteer Service. The volunteer program would allow people to serve for 2-3 years to work on global ecological projects, emergency response teams, or peacekeeping teams (possibly to get credit for obligatory national service).
- *Manage the interrelationships* of nations to coordinate trade according to higher standards. Through the Office of Licensing, satellites, space and the electromagnetic spectrum would be licensed, for example.
- Create a International Courts, such as Rights, Interests, Environment, and Tax Courts.
- Create a Commission on the Earth, as a trusteeship over global commons, to coordinate surveys, inventories and monitoring, for everything from oceans to space. Create ecological goals for restoration of global systems, especially the ocean and forests.
- Set up Commissions on Culture & Religion, which would implement cultural goals to protect the health of human communities, and support educational activities.

- Create a set-aside account to deal with global catastrophes, as well as geological and astronomical threats to the planet and to nations.
- Create a Planning Commission for the Earth, which would protect hotspots, critical and common areas, as well as restore large areas.

These powers would allow the UN to act quickly to protect the planet and to bring peace to the many cultures and nations. People could suggest further changes.

Refine the Responsibilities of a Nation

Every nation has certain responsibilities to itself, to its environment, and to its human members. These include:

- Define & secure good borders, to keep identity, to keep vital processes inside, but allow people, energy, and materials to come inside.
- Balance isolation & connectivity, to preserve the advantages of differentiation and invention, and to limit trade and participation.
- Balance Immigration with Emigration, to balance certain skills and subpopulations with its overall goals, depending on their trade specializations or desires.
- Secure Treaties with Neighbors, to normalize relationships with its close or regional neighbors. Such official agreements would spell out the extents of exchanges.
- Set Standards for requirements, tariffs, protections, and for emissions and other things, such as nature and ecosystems. A nation should also set the official standards for those things necessary for political interactions or economic production.
- Establish Rights for People and Nature, where Humanity has the right to coexist in healthy diverse conditions. Basic expectations are: Equality of opportunity; jobs; security; share in luxury; preservation of civil liberties; equality before the law, and, enjoyment of the fruits of progress in a good standard of living.
- Represent All People. A nation has to give equal representation to everyone in its borders. Although distinctions will be made between citizens, noncitizens, visitors, and tourists.
- Amend Constitution to control Corporations & Groups to reflect new international standards and agreements, but to reflect the new responsibilities of nations themselves, and of their constituents, especially powerful groups or large corporations, which will no longer be legally recognized as human individuals.
- Encourage Small Businesses, whose flexibility has provided lessons for big businesses and which contribute significantly to the economy of a nation, in innovation, in adaptability, and in job creation for women

and minorities, as well as to distressed or depressed areas.

- Create Long-term Ecological Planning to reflect the importance of an ecological perspective, to know what habitats and resources are within a nation, to be able to delineate them, and to evaluate their health by monitoring.
- Protect, as well as conserve and limit use of Resources, especially non-renewable or slowly renewable ones. This may be temporary, for future resources or use, or permanent, for areas critical to regional and global cycles. Promote appropriate technology.
- Plan for Human and Cultural Needs. Encourage vernacular design, as Ivan Illich suggests, to express a shared heritage in patterns of construction of shelter, and foster ecological design as Sym Van Der Ryn and others recommend.
- Balance Total Budgets, especially new taxes for equalization of incomes and for internalizing national costs for elements, air and water (Use Taxes), for oil, forests, extinctions (Loss Taxes), for pollution, cigarettes, alcohol, drugs, financial speculation, heroic income & possessions (Adjustment Taxes), for voting, media, business, marriage, reproduction (discussed by J. Javits & B. Packwood), driving, wildlife collection, and weapons (Licenses), for roads and tourism (fees).
- Require a Two-Three Year Service, to carry out many of the plans and operations of the government, as civil or police service (similar to national guard).
- Operate national government, with its leaders, representatives and departments. In addition to creating jobs, the government would issue vouchers for basic income and health care, as well as for education. Vouchers for elements and water may be issued, also.

The UN might have to power to enforce these responsibilities, which would not be optional or evadable.

Figure. United Nations, NY (Credit: UN).

Humanity would have Basic Rights

Civil and political rights are sometimes divided into negative and positive rights. Negative rights, which follow mainly from the Anglo-American legal tradition, denote actions that a government should not take. These include right to life and security of person; freedom from slavery; equality before the law and due process under the rule of law; freedom of movement; and freedoms of speech, religion and assembly. Positive rights follow mainly from the Continental European legal tradition, which denotes rights that the state is obliged to protect and provide. Examples include: The rights to education, to a livelihood, and to legal equality. Positive rights have been codified in the Universal Declaration of Human Rights and in many twentieth century national constitutions.

Human rights can also be based on the 'natural' moral order described by religious precepts. Religious societies justify human rights through religious arguments. For example, liberal movements within Islam have tried to use the Qur'an to support human rights in a Muslim context. Some basic universal rights can be identified.

- *Right to Healthy Environment, Air and Water.* Principle 1 of the Rio Declaration states that human beings are "entitled to a healthy and productive life in harmony with nature." This statement falls short of recognizing a healthy environment as a basic human right. People need to be assured of having clean air, clean water, and healthy ecosystems.
- *Right to be Secure.* The right to be secure can mean being free from invasion of the home. It can mean access to wilderness or to land where you can provide for yourself.
- *Right to Opportunity for a Home.* The right is in the opportunity, not in the ownership.
- *Right to Opportunity to Minimum Work.* Every person should have the right to work and to receive a living wage for their work. A change in the laws, or in the Constitution, could provide every citizen with this opportunity. Most people want meaningful work. People want to contribute to their own well-being, as well as to that of their family, community and nation. It is in the common interest of the community and nation that people who work should not be poor or dependent on others for support.

Other social and personal rights might be considered tradeable rights, such as the right to reproduce or the right to a higher level of luxury. If certain things were rationed, such as energy or consumption levels, people could choose to give up some things for certain kinds of more luxury, such as air travel to expensive island resorts.

Nature would have Basic Rights

Rights seem to follow the expansion of the sphere of ethics, as formal statements of intuitive knowledge. Thus, rights for nature are now being considered. Paul Shepard says the argument is not new, and that its application is ambiguous because 'unlimited rights' will conflict with human interest. But, he makes two assumptions: That human interests are not ambiguous—they are—and that animals will be granted unlimited rights—they will not.

The strongest argument for rights is the interrelatedness of natural and human communities, which is the basis for assigning rights to nature. Life is more than competition; it involves cooperation and play. Rights are formal rules for living together. It would be foolish not to assign rights to animals, plants, and the earth because of contractual formalities.

Humanity has taken its own opportunities. These opportunities have been codified for centuries as rights. Now, we must allow other beings equal opportunities. The interrelatedness of life dictates the interrelatedness of rights. And these rights are necessary to the integrity of the whole planet. Humanity developed in a community of animals and plants, as part of the same tree of life. The quality of human life has always depended on the quality of animal life. Animals have sensations and feelings, as important to them as ours are to us.

Furthermore, the extension of rights to animals and plants does not deny any traditional human rights. Animals should be accorded higher moral regard and legal standing to reflect the intrinsic worth afforded by their existence and sentience. Current welfare laws to conserve species and to guarantee humane treatment in research, transportation, and slaughter indicate a growing concern among people. A new ethic can keep animals free from human intervention, prejudice, or overuse. The intrinsic worth of animals is independent of instrumental values imposed by us.

One problem with the current legal system is that all nonhuman beings are given the status of inferior human beings or legal incompetents, thus keeping humans in a guardian role. A new legal category is needed that would respect the existence, competence, and excellence of natural beings. Christopher Stone recognizes that the judicial system has granted rights to a variety of inanimate holders, trusts, corporations, and nations, for instance. The legal system already operates with fictions, so the extension to natural entities should not present a problem. Two basic rights for nonhuman beings in their species are: The *Space to Exist & Opportunity to Flourish* and *Freedom from Premature Death Extinction & Suffering.* Every species has to be allowed the opportunity to live, even species that we fear or dislike, such as sharks or viruses. We do not know how these species contribute to the whole process of nature. Giving other species opportunities does not mean sacrificing any human needs, just limiting human influence and interference to a designated percentage of the earth, perhaps 40 or 45 percent. In our control of artificial areas, which include many wild species, we can imitate the process of ecosystems by allowing birds, bats, and other animals the opportunity to distribute seeds and energy to other areas or to access their prey, which may be our 'pests.'

Animals do not need to be saved from natural death, which is a great regulator of life, but from unnecessary suffering, experimentation, and premature extinction. The world would not be a better place without sharks, silverfish, rats, cockroaches, or hyenas. They need their own places, where they can take their own opportunities, live or die. The places, entire ecosystems, need to be saved. If we diminish variety in nature, we debase its stability and wholeness, which we need to survive.

Yawn, Another Earth Day

Human consciousness of our effects on the planet and on each other has gradually increased, as shown by recycling programs and by celebrations such as earth day, which has been getting larger every year, as a celebration. But, celebrations did not seem to have lasting effects. Environmental deterioration has worsened. Levels of consumption have increased; populations have increased.

This year's earth day celebration is over. What were we celebrating? That we are going to save the earth or maybe just still think about it? Perhaps we were celebrating our intention to go on a material diet or an opportunity to spend money on t-shirts and buttons. Perhaps earth day is a new springtime variation of a new year's resolution—a temporary awareness, a limited intent, and a reason to party before business as usual. Or, perhaps it is a modern penance that allows us to buy a place in heaven by promising to save the earth with small tokens.

The token changes and vows help, but are they enough? Will a little conservation avoid a great human disaster? Are these easy remedies reminiscent of medical cures for diseases and problems, such as smoking, overeating and stress, that could be avoided by simple denial. The implication is that a few small things, like recycling bottles, will save the earth—that ozone depletion, rainforest destruction, population growth, and painful inequities will somehow be corrected automatically, as governments and industries continue as before.

We have been told that saving the earth starts in the home. No wonder corporations give their blessings to this event—most pollution and waste is industrial and agricultural! Which issues have higher priority? Deadly local ones, such as toxic waste dumps or topsoil loss, or deadly global ones, such as greenhouse gases or chemical runoff? Is an alarm justified, or is caution enough? Are we too lazy to follow through with the effort that we already celebrated? Are we too cheap to deduct a required percentage from profits to pay the real environmental costs? Where is the will and the vision necessary to really make radical changes? We have, in fact, taken the easy way at every branch. We have assumed that corporations will choose the proper path of production and regulate their pollution. Yet, we know they don't.

We have wasted forty years attacking the symptoms and not their technological or social origins. We must acknowledge the failure of our remedial efforts, and our failure to address the flaws of our designs and ideologies.

I do not want to drive fewer miles in a high-powered gas guzzler—I want to travel by train; I want my radio shipped by train and not truck. I do not want my vegetables shipped at all, but grown locally. I do not want farmers to do slightly less aerial spraying of fertilizers and biocides; I want organic produce. I do not want to recycle aluminum and plastic, I want returnable glass containers. I do not want safer coal-burning centralized power plants, I want local solar power. I do not want to give more money to the homeless; I want my tax money to help them build and keep their own apartments or homes. If industries cannot help me with what I want, then I want my government to channel them, tax them and regulate them. And, if my government cannot do that, I want to encourage a change in government and support better candidates.

I want my representatives to ban CFCs, to ban burning, to tax non-recyclables—and if they do not, then I will run for office myself. We do not have time to look at all the information that we have collected, or to convert it to knowledge. We never have had time. We cannot connect with all the information flows. So, we will have to act as if the information we have is enough for wise decisions.

The earth does not need to be saved or healed, as if we could do either. The ways of life that we remember and prefer, the places that depend on other species and natural processes—these can be saved. Our own divided minds, that let the poor be enslaved by the wealthy, that let 'good' animals be domesticated and 'bad' animals be eradicated, can be healed. The sacrifices will have to be great; the changes will have to be radical. But, the celebrations will be meaningful only then.

Figure. Earth Day poster
(credit: Peter Max).

Reintroduce Whales Back to Scandinavia

Waiting on a delayed colleague in northern Norway, I sat on the dock and watched fishing boats come back for the day. Many seemed to be empty. When one unloaded some shrimp and the men were sorting them, I strolled over and asked how it was going. The story was frustrating to hear. There were no fish at all close to the national limits. But, shrimp could be caught. I bought some for dinner that night, still alive but barely moving.

My colleague arrived and we went to his house for dinner. I asked about the fish and heard a tired story of overfishing. I asked what whales were native to the North Sea. He quickly said that minke, humpback, sei, and orca whales used to be present in the fjords and the Norwegian Sea. Algae blooms provide nutrition for zooplankton and pelagic crustaceans— copepods, amphipods, shrimp, which support whales, as well as seals and birds.

The Norwegians have hunted whales for thousands of years, considering them just as food animals, and now research subjects that could be later canned and eaten. The use of maximum sustainable yield in the wildlife management of whales has resulted in the degradation of the populations. The destruction of large species, like whales, has a dramatic effect on ecosystems. When whales were mostly eliminated from the waters in Antarctica, the krill populations crashed, instead of expanded; whales provided nutrients and an environment that enhanced the krill. The destruction of microbes, which generate oxygen and recycle nutrients, has a critical impact on the entire food web. Many unowned species in the global commons, such as blue whales, are hunted to extinction.

Some managers, like whalers, are far worse. They do not try to manage for a continued maximum yield; they try to maximize the economic value of a resource, within economic time limits, in spite of an awareness of extinction—the rape of one "resource" provides the capital for the rape of the next.

Perhaps we should not argue that things have value in the human system. Let us just respect the ultrahuman system: Whales have whale value. Wolves are not efficient at binding nitrogen; neither are humans. Lichen are poor predators, but they break apart rock better than big-horns. The world would not be a better place without slugs or sharks. Their existence has value; they have functions that intersect with other members of the food web and contribute to the health of the whole system.

My colleague helped me write an appeal in decent Norwegian to the national parliament. We agreed that I would emphasize the positive aspects of reintroduction and the benefits to the sea, land and people (I would not criticize them for stupidly overkilling whales and fish). I started by saying that by restoring whales to their waters, after so many decades, Norway could take pride in restoring a healthy ecosystem that people, including tourists, could really see and admire. And, the resulting richness might allow some small amount of exploitation of many other species, especially of fish. Some species, such as whales, might not be used at all, but that was an ethical decision that only the Norwegian people could make. We delivered the letter, and spent the next month presenting the argument in environmental classes at the university. We thought that Norwegians might adopt this suggestion and appreciate their whales without taking even a small fraction, which might put pressure on their reproduction and health. There was no answer. I want so badly to see whales charge into the fjords someday, feeding, spouting and playing as people watch.

We Need New Words!

So many words are useless now for communication: Sustainability, nation, wilderness, progress, democracy, constitution, rights, religion, and so on. Using new words might result in fresh perspectives. Consider *ecodeontics*, from the Greek word fragments meaning 'bind to the house.' Thus, ecodeontics is a way of binding power or managing the invisible in a pattern of ecological systems; it might be better then 'religion.' Religions have been attempts to understand or control the world, either by understanding the invisible or by having spiritual beings intercede. The stories of religions concern events that are deeply meaningful to the listeners. This helps bind the group, also. Religion may help control disruptive forces, especially things about distribution and power. Religion coerces people into a social contract. Religion and story-telling can reduce the variability of individuals in a group. But, this might increase variability between religious groups. Ecodeontics might avoid the connotations of infallibility or violence associated with traditional religions.

We need new words to describe the holistic changes proposed for our very civilization. The equal apportionment of 'resources' to all cooperating participants in the global commons could be identified with the new word '*Koinomics,*' which would be supported by the theory and practice of recognizing and honoring the legacy of the entire planet that hosts its legatees as tenants, which would be represented by the word '*Legatism,*' which would be supported by the 'rule' of all beings, identified as a '*Panocracy,*' although in the human legal system, humans represent the interests of all other beings, much as they are starting to do now. This reapportionment would be enhanced by the wisdom of harmony, called by the word '*Harmosophy,*' and by the drawing and making of ecological zones, through '*Zonagraphy*,' which emphasizes the relative isolation of wild and artificial areas. This reapportionment of 'resources' that human communities have already claimed, as well as of resources that have been badly distributed as a result of theft or violence, may cause some degree of discomfort for wealthier people, but that is minimal compared to the suffering and death under the current system, which encourages overconsumption and immoral differences in the distribution of wealth.

If we think of cultural ideas as *memes* that can be designed, then we can approach them as designers not simply as users. For instance, the memes we would consider range from culture, and its technology, agriculture, domestication, cities, economics, and politics, to wilderness and nature. All of these are human constructs that have boundaries and centers, but all of them are also separate from the human sphere, although they interpenetrate.

A unit of dwelling could be called a '*deme*.' William Irwin Thompson used this word when he advocated a meta-industrial village he called a deme, but the word could be applied to any species, from termites to wolves, and to any size unit, from villages to cities. The deme is a useful unit of design. A unit of wilderness could be called a '*wene*.' This would be a regional ecological complex, similar to a '*rene*,' which could be a unit of ecosystem renewal. The pathways through and between ecosystems could be described in terms of '*venes*' (don't laugh, they rhyme). A human ethnic or cultural group could be represented as a '*cene*,' which would not have the cultural baggage of nation or republic. The specific goals of a group could be a series of '*zenes*.' Which would be accomplished through specific actions, '*phenes*.' We might even designate units to represent limits ('*lene*') and principles ('*tene*'), in order to discuss the importance and commonalities of these things. Design could combine these units into new patterns that might be more flexible or resilient. Thus, renes could be combined to make up a restored wene, within a matrix of demes and cenes. On the other hand, maybe inventing special new words is silly; maybe we just need to understand the older meanings of words and use them properly.

Figure. Fun with Words (Credit: aphaDictionary).

Divide the Planet Into Zones & Save Wilderness

We have been entertaining a total reconsideration of the current pattern of technologies, cultures, value systems, and behaviors, evolving into a low-profile technological ethic suitable for a renewal of ecological understanding. The purpose is to preserve those parts of the earth that are needed for renewing the holecosystem, which contains habitats for the billions of animals, plants and living beings that are part of the planet. It is to allow fair use of that part that is human, and for human equality in opportunity and worth. It is the demand for a margin from catastrophe, so that if humanity is unable to live peaceably, the rest of the earth will not become extinct.

The United Nations could divide the earth into zones for preservation, conservation, domestication, and human communities. Human activities would be limited to specific zones and, within those, a global authority controls all air, water and land use under complete sovereignty. The UN would calculate an optimum human population for each nation based on a calculation of net community productivity on arable land through traditional agriculture. Common planetary resources would be assigned according to the optimum population figure. The development of each nation would be regulated by the UN through charters. Self-reliant nations would decide their own appropriate technology, crops and institutions based on traditional values and be responsible for the ecological education of their constituents. Residents of nations would have equal rights and work opportunities, and have the responsibility to participate in government and to make good places. This division would be made through a presentation of properties, principles, standards, and practices that could be combined to address challenges and problems facing our species to make good places on earth.

Properties are characteristic qualities of a thing that distinguish unique individuals, systems, or patterns; Gregory Bateson calls them differences that make a difference. Properties are also essential or accidental. An essential property is one that a thing possesses regardless of relationships with other things. If an essential property is lost, then the thing becomes another species; an accidental is one that makes less difference to essential properties. Principles are fundamental rules or laws, derived from and based on the properties of the surrounding systems, that we can use to create images or models to meet stated objectives, that is, the goals towards which our action is directed, e.g., a healthy ecosystem or a balanced productive society. Principles unify our images. Standards are models or examples of quality or value, established by authority or mutual consent, that can be repeated as procedures. Standards are established from principles. Practices, as actions or procedures, are regular operations based on standards and principles. Principles match properties to our standards and practices to maintain the system favorable to a community of all beings.

For example, one property of the planet is its wildness. The corresponding principle is that the planet is self-making and self-ordering, without requiring human control and management. Our objective for the planet is to allow the wild process to continue. We can set standards that are likely to keep the planet wild, stable and enjoyable: Limit our exploitation to less than fifty percent of the surface area; use appropriate techniques to minimize damage to wild systems; or, calculate optimum regional populations based on ecological and cultural carrying capacity. We can match our practices to those standards. For example, we could restore areas that have been damaged; rework or concentrate urban areas rather than converting wild systems; or, limit our population to one that is flexible to avoid catastrophes and enjoy different standards of living and luxury.

Reimagine Agriculture

Many of our problems can be traced to agriculture, which was an adaptation to the climate chaos after the last ice age. It seemed to be a good way to take advantage of stands of wild cereals, choosing the best and planting them. Using them every year. It started as a status symbol, and finished as a trap. It required more work, but it allowed more children, which were useful to perform more work. It offered more calories but fewer vitamins. People became less healthy and shorter. There were advantages, such as a storable surplus, but that led to inequities between people and conflicts. It lead to writing to keep track of where the harvest went, but writing led to changes in power between the sexes and between groups of people. Land was altered for crops then enclosed with boundaries. Technology supported large plantations of single crops, using slave labor then machine labor. Crops became commercial and people had to buy them. Preparing soil resulted in erosion. Poor soils required fertilizers, which hurt many plants. Problems with pests required pesticides, which killed many good species. Cleaning up fertilizers and pesticides took many millions of dollars. Agriculture required more technology, which separated people from the land and food.

Agriculture has become a trap, allowing populations to expand and requiring it to expand; we are afraid to abandon it, lest people starve. The genetics of crops was continuously narrowed as the number of crops narrowed. Profits were maximized as diversity was narrowed. Crops were pushed into inappropriate ecosystems, bolstered by demands for borrowed ancient water and cheap fossil fuels. Wild ecosystems were converted to artificial ones, until 30% of net primary production of the planet was diverted to suppoprt humanity (in 1974; 40% by 1992; 56% by 2001). The nutritional value of foods declined and medical costs increased.

Agriculture has been a primitive way to use the nutrients and energy from an immature state of succession to maximize growth. We have to make it a mature system immediately. We can do that with a more permanent agriculture, suggested by Permaculture. We can do that by intercropping, suggested by Fukuoka and others. We can do that by planting perennial grasses in a mature system, suggested by the Land Institute. Agriculture can be made a part of cities as Soleri and others have suggested.

Stealing from the future is drawdown. Humans should learn to live within the carrying capacity of the ecosystem they live in, relying on renewable resources consumed at sustained rates. Much of the disruption and damage from agriculture, as well as human suffering and loss, could be avoided if agriculture was modeled after mature ecosystems rather than pioneer stages. Civilization would be more complex, although smaller.

Stop Fishing!

Ocean fishing has provided protein for people ever since they learned to fish. The current high demand for fish, however, is causing problems for natural fish populations. As natural fisheries are being depleted, and collapse, industrial fish harvests are becoming more common and destructive at a time when we are being embarrassed by our ignorance of the ocean.

Although 100 new fish species are being identified per year, probably 50 others are being driven to extinction. Coral reefs are mined; 50 percent of the coral reefs damaged and perhaps 20 percent destroyed. It is just a fish filet for dinner, what is the harm? And the answer is very little harm at all—it's just that on a massive scale, little sins can kill numerous species and destroy large habitats. And, few of us see the results, just fewer fish or squid. The Gulf of Mexico is dead for hundreds of square miles. Why? Garbage, phosphate dumping? Who knows what combination exactly, but it killed the fish, crabs, and almost everything except red tide bacteria.

As the good fish disappear, we move on to the next best fish and finally to junk fish that can be ground up into patties for dogfood, or fish cakes. Even a modern, balanced exploitation may destroy fisheries. Currently, many resource managers espouse the ideas of equilibrium maintenance and maximum sustainable yield. These ideas are poor guides to management. By trying to maintain habitats in equilibrium, we often set them up for catastrophic decline. The use of maximum sustainable yield in wildlife management has resulted in the degradation of the populations involved, salmon, for instance.

Even the ocean is not invulnerable to human impacts. The most productive areas, which are close to shore, are being polluted, overfished and destroyed. Estuaries and coastal wetlands are being destroyed. Overfishing is depriving people of millions of tons of seafood in the future. Overfishing is destroying the fisheries' support systems. In the US, losses to fisheries from shore "improvement" and degradation cost $86 million a year. Estimates of potential catches in the ocean are characterized by irrational guesses. Almost all tasty species have been overfished to rarity or extinction. Less desirable species are taken in greater quantities for animal feed and fertilizer; over 36% of the global catch goes for animal feed.

Aquaculture is touted as the solution to the 'emptiness' of the ocean. However, it has many disadvantages. The costs are high, from the structure near the shore to the impacts, such as pollution and disease generation, on the surrounding aquatic ecosystems. Fish are raised and kept in tanks for their entire lives. They are born in a hatching tank, then transferred to a tank and crowded together with other fish. Overcrowding creates unsanitary conditions for the fish, from excrement, chemicals, and antibiotics. But, fish, like chickens, cattle, and pigs, are capable of suffering. Another concern is the spread of diseases. Wild populations of fish are affected by aquaculture in other ways as well. The spread of parasites, such as sea lice, from farmed fish to wild schools, might occur. Finally, there may be genetic mixing between wild fish and their escaped farm fish. People argue that environmentally sustainable ocean aquaculture is possible and necessary, if the right species and techniques are used. Open-ocean fish farming might work, but it is remote from observers, and may be neglected. Until these problems are solved, it might be best to discontinue aquaculture.

Possibly ocean fishing could continue, if the take was reduced by 50-90 percent. The Polynesians had a taboo system that limited the number and kinds of fish taken. The Shaman indicated which fish needed to recover and if you were caught violating the taboo, you could be killed. We need a global shaman to say that the Ocean, and many of the seas, such as the Mediterranean, are taboo for fishing for several generations. It might not be popular, but we could survive on other foods for a while, and it would be a meaningful sacrifice.

Restore Trashed Ecosystems

Many systems that have been degraded or overused to collapse. We have been converting complex wild ecosystems into simple artificial ones. So, restoring the wild ones is a good idea. But, often we do not know the composition of the old systems, since no one bothered to take an inventory of plants or animals—or studied the relations and cycles of the parts of the systems. We need to create restoration projects to ensure biotic survival. We could try with informed guesses to set aside systems and add equivalent exotic species back in. When we do know the species, they may not be available. They will require intensive management, even for sets of species in small areas.

So, what should we do? With the recent emphasis of ecosystem services—and that is a human-centered concept, since we are the recipients of those services—and their approximate value or costs, we could create hybrid systems, using either novel or invasive species to provide similar services. That might be a start, since if the systems worked on any level, they might then develop into fully functioning systems composed of interactive species. Possibly remaining native species could reestablish themselves over time. If the system is restored cleverly, it may not need further intervention.

There is little information on the restoration or preservation of 'near-natural' ecosystems, those that are primarily native, and not subject to major change. But, these systems might be restored—not to an original state since some grass species are missing—but it could be rehabilitated into a functioning system, if human exploitation was limited in the future.

Most restoration areas are experimentally restored ecosystems that had been severely reduced or disturbed by human activities or natural events, for example, Lake Shagawa or the Tall Grass Prairie in central North America. Because they are so minimal in extent, now, they might be called neopoetic areas. Some areas have been started by human activities, but saved as self-regenerating systems and not significantly built up or interfered with. An example is California grasslands, which have been changed with Mediterranean plants, e.g., wild oats, wild mustard, radishes, and wild fennel.

Science can start to plan for future ecosystems, which may be less beneficial for humanity. Many of these will be restored or assembled from exotics, reintroduced natives, or engineered species. Science needs to save as much genetic history as possible for that effort of restoration; that may mean using cryogenic techniques to suspend material from rare or endangered species. Science has to anticipate the loss of species and the massive impacts on ecosystems that result from human technological and economic impacts. It has to anticipate ways to ameliorate those impacts with restoration.

Places cannot be completely preserved, without cutting their vital connections and making them lifeless. Places cannot be restored to some Arcadian fantasy, without severely limiting their movement and development. Places cannot be created using a machine metaphor that boasts the replaceability of anything. We have to work on the ground, carefully selecting native or near-native species to mix with other species selected from similar niches or species that could offer similar functions in the system. Then we can wait and see if the neopoetic systems create the same cycles and outputs as the real systems we transformed or degraded. It is difficult to determine the best practice to use, especially when applied to long-lived ecosystems. We have to act as if we knew or as if we were wise.

Show Reverence for Life

The physician Albert Schweitzer noted that ethical thought had been developing since prehuman history, and it culminated in the principle of 'reverence for life.' During the Renaissance, love of man was the virtue above all. Philosophies of the 18th century constructed a natural ethics where reason supported the commandment of love. Schweitzer believed that the ethics of Jesus, reinforced by reason, lead to the 'reverence for life,' whose edict was the rule of universal love. He considered that any thoughtful person could not help but enlarge the scope of ethical activity until it included all nonhuman life. Schweitzer noted that during the evolution of humanity, the circle of responsibilities had gradually widened, beginning with family, then tribe, nation, and humanity—working toward all of life. However, he concluded that nature had no reverence for life. It produced life a thousand-fold in the most meaningful way and then destroyed it a thousand-fold in a meaningless way. The most precious form could be sacrificed to the lowest. Sympathetic concern toward all the wills to live was the basis of ethics. Reverence for life was the greatest commandment in its most elementary form.

Schweitzer stated that we must struggle to wipe out antihuman traditions and inhuman emotions, that we must struggle against our own insensitivity. It is inevitable that we kill some things, unknowingly or to survive, but we must never come to take killing lightly, plucking flowers and squashing ants indiscriminately; we must not become thoughtless and blind. He urged a new renaissance, a spiritual renewal, an adventure where ethics must stretch out not simply to mankind but to all creation, and especially to all life.

Although Schweitzer showed disdain for primitive religion and animism, the belief in souls and the spiritual being of trees and animals is embodied in the reverence for life—to be one with the animal you hunt or with the corn you grow, to realize that it becomes part of you, living through you. However, for Schweitzer, nature was a cruel drama of the will-to-live divided against itself. He considered the enormous mortality in nature to be an embodiment of evil. He was most concerned with the protection of animals useful to humanity, rather than the preservation of all animals. He himself killed wild predators to save domestic goats.

The fact of life entails death, however. The dying process is integral in nature for the continuity and renewal of life. True reverence for life has to entail reverence for death, since life and death are inseparable. No pattern can survive death, when death is the destruction of individual patterns. All life is sacred, but this can never be a reason for not killing, because that is how lives are sustained. Since life is of the utmost importance to the living, it should only be taken in sorrow, used and shaped with respect, and experienced with awe, for underneath it is still unfathomable mystery. M. W. Fox judged that Schweitzer's ethic was flawed by having no ecological ethic.

An ecological ethics is based on attributes of ecosystems and on human compliance with ecological laws. Science could demand an ethic directed to the preservation of life in its mosaic setting. But, only a religiously conceived ethic has done so. And, Schweitzer's reverence for life is the only one visible in Western world. His reverence for life principle acquires a new aspect when it is restored to ontologically firm ground. The world becomes a synthesis acquired from values with the mysticism of religion, characterized by love, compassion and the reverence for all things. As the circle of ethics was enlarged to include the realm of all living things, it could be stretched to include all things, animate and inanimate.

Make Education Ecological

Modern education allows individuals to languish in an informational wasteland. An education based on the aesthetic humanism of Frederich Schiller could lead them out, to a place within nature. It could offer a new perspective of humanity in the total field on nature and define a balanced relationships with other species. At first, Schiller believed that human society could be improved by political means. But, after studies on the Thirty Years War in Europe, he became skeptical of the ability of politics to create a peaceful society. He came to consider a work on art as historical proof that art could achieve what violence and law could not: Art could educate and liberate the individuals of society in a gradual and peaceful process. In spite of the cultural forces dominant at any moment, an individual had the potential to determine a different course of action. Unlike classical humanism, which was shackled to one interpretation of the past, the aesthetic humanism of Schiller was open to the possibility of novelty.

Schiller judged society to be violent and selfish due to an imbalance between animal and rational drives; civilization exacerbated the problem with its own imbalances, for example, fragmentation and the unnatural channeling of energy. But, society and civilization could be reformed through education.

An education based on Schiller's ideas could present a whole image of humanity within nature. It could confront the past without the baggage of sentiment and the future without the paralysis of dread. Furthermore, the appreciation of the differences of other cultures could allow human beings to enlarge their experience and identities. Art could broaden the mental worlds of observers and encourage tolerance and wonder. Education in aesthetic humanism embraces three concepts: play, liberation, and community.

Play is the method of learning for most juvenile animals and a means of enjoyment for many adult animals. For humans, play is imaginative experience; even science and philosophy are forms of play, attempts to solve the puzzles of existence.

Liberation requires a larger perception and larger concept of rationality. Liberation means an end to prejudice or discrimination based on arbitrary characteristics. The liberation of nature and ultrahuman beings is inseparable from human liberation.

Community is the context for education; it is the means for communities to continue. As Plotinus and Novalis recognized, education has an outward, social and civil aspect as well as an inward, personal and self-revealing aspect.

Education has at least four ends: The appreciation of the richness of nature; the comprehension of human existence; the understanding of the nature of human society; and training for a position in human society. Centuries later, education has become more universal, but its goal, the well-rounded individual, has been distorted by its fourth aim, training for the economy. To produce wealth for the state and livelihood for the individual, education has become money obsessed. Education, in the second and third aims, has been neglected, since it might limit or contradict its economic obsession. In fact, the first three aims are restrictive to a growing, industrial economy.

With its emphasis on play, liberation, and community, an aesthetic education reduces the emphasis on training and salary and integrates all four ends. A radical, aesthetic education alters and enlarges perception with the selection and presentation of relevant information and forms an ecological consciousness. The survival of human societies depends on consciousness of the global system in its complexity and connectedness. The spirit of humanity depends on the consciousness of its proper relationship with other species and the whole earth.

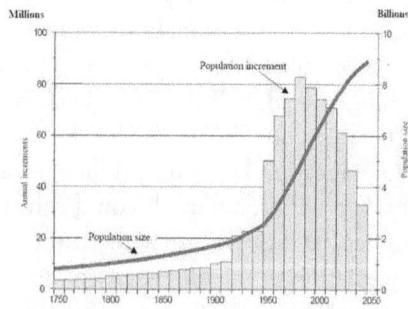

Figure. Long term population growth (Credit: UN).

Identify Gigatrends, Pay Attention!

A number of trends in human society, as well as in ecological systems, are evident. A few of them can be fit into the megatrends identified by John Naisbitt over a decade ago, for larger patterns in society, including the move from industrial society to information society. He considered time frames from two-year horizons to a time frame of six to ten years. The very large, very-long-term trends, such as atmospheric temperature increases, ecosystem conversion, or global deforestation were ignored by him, and by science and economics mostly.

These trends occupy the entire human calendar, as well as the development of the planet. They are gigatrends. Some of them have been noticeable for thousands of years, but nothing has been done to halt them. Environmental factors have shaped the course of human history to a greater extent than has been realized. The decline of Rome demonstrates that ignorance of forest ecology has important consequences. Salt in irrigated soils left by evaporation led to collapses in Mesopotamia. These civilizations were successful before they failed.

Failure after success is tragic. For the Greeks, the operation of tragedy resulted from success taken to great lengths, that is, where successful behavior in one context was applied to all contexts, with negative consequences. For example, humans in moderate numbers were able to take what they needed, such as wild animals, from natural ecosystems without interfering with the processes. With larger populations, humans have modified animal and plant associations in a different way, simplifying patterns of energy and chemical exchange, converting entire systems for agriculture and cities, and solidifying themselves at the end of many food chains. Our success, because of our big brains and tool-using hands, has become self-destructive, as species go extinct and systems collapse.

Some basic gigatrends include: The exponential increase in human population; the exponential impacts of a small percent of consumers; the conversion of habitats; the drawdown of resources; the continued avoidance of and ignorance of carrying capacity, long-term deficits, and the loss of local and regional connections. Wood is an example of a negative trend. Despite wood shortages for over four thousand years, the price of wood is still nowhere the real costs of production, and wood is used for cheap, impermanent goods. These gigatrends are interrelated and complex.

Many of these gigatrends are partly the result of our unconsciousness of large-scale, long-term events, partly the result of out cultural amnesia about things that make us unhappy, and partly the result of our cultivated indifference—doubtless from our remoteness from wild nature, as a result of our tools and the general abstraction of civilization. None of these gigatrends can be reversed until the remoteness is re-educated into participation and attentiveness. There are already a few trends flowing against the tide. Some positive gigatrends include: The adaptation of human cultures and ecosystems over time in Asia, Europe, and parts of Africa, resulting in rich domesticated landscapes; setting areas, for preservation of ecosystem processes, reservation of archaic cultures, and conservation lands, aside from industrial interference; new trends in ecological cities, led by Paolo Soleri and others; and an increase in the scope of ethics, from family, tribe, nation, humanity, to include reverence for all living beings, identified by Albert Schweitzer.

The intent of describing large-scale patterns is to have human patterns fit with observed patterns in nature. Fitting the pattern can lead to both continuity and predictability, and both of these are needed to adapt human activities to natural limits. Thinking we have conquered nature and are omnipotent, we have quit thinking. Satisfied with our comforts, we do not ask enough of ourselves. With these gigatrends possibly ending in tragedy for humanity, we must ask hard questions and start thinking again.

Seek an Optimum, Not a Minimum

Our civilization has become enamored with the idea of the maximum; it is the goal for many kinds of planning and operations. But, in almost all economic things, we try to just aim for a minimum number: Minimum time, minimum space, minimum wages, and so on. We try to maximize profits and minimize costs, for instance, without understanding what either is. Some engineering projects are built with minimum standards, yet in the long run they often cost more in repairs and utilities. We want to save animals but only leave a minimum viable population or species space; we do not want to set aside space for the numbers that may be needed for keeping healthy ecosystems necessary to keep global cycles running. We figure as long as nonhuman life has minimum conditions it will support us with food and services. We only want to be bound by minimum standards of behavior.

We are creating a revolution in minimalness, a rush towards cheapness that characterizes most of our industrial production, from clothing and furniture to automobiles and electronics, a drive to the lowest cost, a barebones strategy that results in only bare bones. The effects take a toll at every level. We offer minimum care for animals, especially food animals like chickens or cattle, knowing that they will be killed for food soon. In general, we pay minimal attention to health at all, for that of an ecosystem or our own. Is this revolution aimed to benefit all, or only a few? Or do we not think about the real consequences of it? This trend may result in a human being living in the minimum space, 7 feet long by 2 feet by 2 feet; food, beverages and water could be piped in, a two foot video screen on the top could be used for movies or games controlled by finger keys; and, at death the entire unit could be recycled or destroyed, the thoughts of and memory of that person kept in a data base, minimal of course.

Is this a natural thing? In some senses, nature does try to maximize or minimize certain conditions. For instance, a bubble minimizes surface tension by assuming a spherical shape that contains a maximum volume. But, complex patterns have many different kinds of limits. So, it would be difficult to calculate a minimum, or a maximum or an optimum, for the forest cover for a tropical watershed. Just a minimum leaves no room for change or flexibility. And, ecosystems, like human patterns, need flexibility for the capacity to respond to environmental conditions. In fact, the functional biological requirements are rather minimal for humans, being only food, clothing, shelter, and reproduction. Yet, we always seem to strive for choice and luxury in these items, and so much more in travel and entertainment, to mention just two.

What should we strive for? An optimum? What would happen if we did? Would it ruin the idea of market economics? Perhaps we could try. We could easily design our economic systems to work toward optima. Science and economics have identified various minima and maxima, and could identify optima. Our values would be more in line with optima. If we tried it, and found it disappointing, we could always rush back to our minimum-value civilization.

Calculate an Optimum Global Human Population

A number of recent studies have suggested that the human population of the earth could be much larger that it is now; a few concluded that it would have to be much lower. This experiment suggests ways to calculate an optimum global human population, using a deductive, synthetic, conceptual model based on data generated from research on net primary productivity (NPP) and net community productivity (NCP).

Many theorists have estimated maximum sustainable populations for humanity. K. DeWit (1967), for example, estimated the maximum human population at 1 trillion, 22 billion. Colin Clark (1967) put the potential population of the earth at 47 billion at an American standard (or 157 billion on a Japanese standard). Scientists who considered additional requirements tended to arrive at lower potential numbers. H. R. Hulet (1970) suggested that the earth could support an optimum population of 1.2 billion based on plant production. Arthur Westing (1980) estimated a maximum global carrying capacity of 1.5 to 3.9 billion. Eugene Odum (1970) suggested using land area as a measure of human carrying capacity. Extrapolating from his calculations for the US State of Georgia, an optimum global population would be 3.9 billion. Samuel Eyre attempted to describe the wealth of nations in terms of the productivity of ecosystems, specifically Net Primary Productivity (NPP), with nutrition equivalents in NPP for mineral resources and with inorganic assets. His calculation was just over 2 billion.

It is possible to calculate a population using NCP instead of NPP, however. This model is concerned only with the values of natural productivities. Domestic lands were assigned productivities based on those of the original wild vegetation (which in fact is usually higher). Land currently used for agricultural practices was divided into original vegetative states and productivities were added to calculations from wild lands. In practice, crop lands are heavily subsidized by fossil fuels, fertilizers, and pesticides, allowing them to produce 10 or more times the food calories, and thus support a much higher population—how sustainable a population depends on whether renewable energy can replace fossil fuels. When NCP is used to calculate a maximum population, by adding the calculations for each vegetational unit of the earth, the measurements are consistent with previous low figures—1.08 billion—after the same percentages of inedibility and waste are subtracted.

The population of 1.08 billion is almost identical to a flat 1 percent rate of the NPP, subjected to the same loss percentages. The use of increased animal protein would reduce the number correspondingly. This availability factor is on the order of 75 percent. The figure, 1.08 billion, is a maximum. *An optimum population, arrived at by a 50 percent rule, is 540 million.* This would insure against problems due to fluctuations in productivity. Richard Watson arrived at a figure of 500 million, based on American levels of consumption with the present industrial production. Daniel Kozlovsky intuitively estimated 500 million also as an equilibrium population. Arne Naess suggested an identical figure based on traditional livelihoods and fitness.

There is a maximum carrying capacity for humanity, but it is unknown; it could be less than the current population. For humans, this capacity must include domesticates, as human equivalents, since many domesticates compete for protein consumption. Domestic animals extend the carrying capacity, since many of them consume agricultural wastes or use lands marginal for agriculture, but domestic animals are not as efficient as wild populations.

Technology could expand the maximum carrying capacity to some extent, with higher yield crops and resource substitution, but also it could reduce the capacity with unforeseen side-effects, from the use of pesticides, for example. War and social disorder reduce the ultimate capacity. Furthermore, the capacity decreases as the per capita use of energy and resources increases. Carrying capacity calculations often just consider food energy, but all needs—clothing, shelter, transportation, information generation, aesthetic satisfaction, wilderness (as places for other species), ecosystem functioning—must be included.

There are also minimum viable populations to be considered. For the human species, it is unlikely that the lower limits will be approached in the foreseeable future. There are several lower limits to keep in mind, however, such as ideomass and genetic minimum. Minima could include genetic, 2000 people, and ideomass, 600 million. Maxima for wilderness might be 1.4 billion; social advantage 2 billion. Many theorists worry that a reduction in human natality could cause humanity to stagnate. Natality is the birth of new ideas, also. J. B. Calhoun uses ideomass to supplement sheer demomass; ideas and culture can create endless possibilities to support a stable or declining population without reducing the quality of interactions.

Human populations are not based on simple mathematical calculations. Sometimes the political aspect of population is of utmost importance, for instance, when the minority groups attempt to outbreed a majority. Calculations of population and carrying capacity are made difficult by emigration and drawdown.

Changing population traditions and growth may be difficult. Slowing growth may be more demanding and require some form of institutional control. Paul Ehrlich has noted that human numbers press against human values, that is, changes in quantity precipitate changes in quality. The quality of life will sink lower long before the limits of food or space are reached. Overpopulation has already blocked many people from desired adventures: pioneering or hunting as a way of life, food and land for everyone, or an airplane in every garage.

Population involves the entire humanity-technology-environment spectrum; it cannot be approached from just one direction. Lower densities of humans will always be able to harmonize more successfully with ecological processes. For the long-term survival of the human species, adaptability to environmental changes is necessary. This requires a wide diversity of gene pools, which is achieved by a relatively large population divided into local, partly isolated groups. And this requires healthy regional ecosystems. The optimum size of the global human population is the sum of optima for local habitats. What is important is not how many people can exist at once, but what kind of life is possible for those who do exist.

Before an optimum human population for the world can be meaningful, other questions must be addressed, for instance: How much land should be left in its native state? Enough to save one of each kind of ecosystem? At what level of luxury should humans live? What are the physical limits of resources? And finally, what is an optimum?—Laboratory studies with rats show that, with a choice of optimum rat environments, some rats will reject the optimum. Is it meaningful to speak of an optimum for all of humanity?

How these questions, and others not asked, are answered determines an optimum human population. In calculating an optimum population within ecosystem restraints, we have to consider minimum wilderness preservation, air and water quality, genetic minima, nonrenewable resources, appropriate technological innovation, the importance of cultural frameworks, adventure, research, beauty, uniqueness, and other intangible experiences—although these dimensions cannot be quantified. The number 540 million people tries to address all these concerns. If human civilizations were based on complex ecosystems, they would be more complex. The complexity encountered in nature may serve as a zen koan-work, to bring us to rational breakdown and to an alternative: Humanity as a self-conscious, self-limiting, poetic species.

Get Down to an Optimum Population

Although we manage other species, we have trouble limiting our own species. Why is that troubling? Does it violate some idea of maximum growth? Or, is there some absolute right to have as many children as we want, regardless of the social and ecological consequences? As a crude comparison, we often limit the weight of our own bodies. We limit the number of televisions sets in each house.

If we accepted an optimum level for human population, could we get there and stay there? That is a complex question. There are many factors leading to an optimum number. We could use available food energy. We could use the raw energy available on the planet surface. Or social advantage or genetic advantage. We could use the ability of the economic network for distribution, although that might be more restrictive. We could use aesthetics, and take into account heroic architecture and travel, both of which would reduce an optimum. If we related population to the minimum of cycles of weather and productivity, and to the frequency and costs of environmental emergencies, that would reduce it. Using ecosystem productivity would reduce the global population to below its current number, but considering current mismanagement and waste, it would be lower. Using energy, 3 kilowatts per person with 6 terawatts for the planet, the population would only be 2 billion, about what it was in 1930, when most people were fed with unsubsidized agriculture, as Daniel Rirdon calculates. Using combinations of other resources, such as aluminum and firewood, the population would be about 1 billion. In each case, with more resource use the population would be lower. And, that does not actually consider most of the impacts on healthy ecosystems and biogeochemical cycles. It seems that while there is no simple answer, the more complex examples point to a much lower population that we have currently.

If we could decide on a number, say 1 .5 billion, how would we get there? Legal enforced limits, such as China's one child policy, do not seem to work. Rational numbers, promoted by conscientious scholars do not seem to work either. What seems to work, as it has for short times in Iran or Denmark for instance, is female education and empowerment. Schooling girls in all nations would probably reduce growth by hundreds of millions a year, and probably cost less than $8 billion per year (less than the monthly cost of recent wars for oil and pride). Equal rights for women in every nation would further reduce it by more hundreds of millions per year. We have been led to believe that reducing population is not only undesirable but also a slow painful process. We should let women decide on the desirability. But, the process might not be slow at all. Alan Weisman calculates that a half-child less average per family per year would reduce the population to 6.2 billion by 2100. A voluntary two-child policy, planet wide, might result in a population of 1.6 billion by 2100. A voluntary one-child policy, under the replacement rate, could bring it under 1.5 billion by 2055 and one billion by 2100. What could voluntary emergency measures and tax incentives do by 2032?

Many people and groups are fearful of a declining population, pointing out that it would be a disaster, especially economically and socially. But, humans are very adaptable. If every family had only one child, most likely nuclear families would be replaced by more extended families that would include cousins and close friends. Economics has always adjusted to sudden population drops, as Europe and China experienced in the Fourteenth-Century. We are immensely adaptable and flexible; we just need to remember that flexibility is more important than efficiency, and adaptability is less important than the capacity for balanced development.

Reduce the Number of Things

People make tools to get food. Then, they make clothing, homes and other things for making life easier. Art, luxuries and sculptures follow. In foraging societies, people often had to carry everything they owned. Horses made that easier, but people still limited in what they owned to a set of ceremonial clothing, hunting/gathering tools, and materials for dance and celebration. With permanent settlements, people acquired heavy houses, furniture and storage boxes (for food, masks and other essentials). With industrialization, people could get much more clothing, as well as works of art, toys, electronic communications and entertainments, jewelry, and machines—for food preparation, tending small plots or yards, and transportation. Many things were specialized, just for one application, such as bottle openers, so each independent household believes one of everything is necessary.

With continued industrial production, many have been able to double, triple, or sextuple many of their belongings, even cars. Some people have to enlarge their homes for the multiplying number of things; others have to rent air-conditioned storage areas for extra things that might be needed in the future. In Japan and parts of Europe and North America, some of the markets are saturated. People just do not want one more television; there is no desire to have 7 or 8 televisions (or 32 for a video wall). How many are too many? How many are useless or superfluous? Perhaps the desire for more things or for new things overwhelms common sense. Perhaps the greed for more things is rooted in the fear of loss or the need for approval. Perhaps possession is related to some competitive behavior.

Too many goods could prove to be uneconomic growth. Many markets are approaching saturation; perhaps others have reached a point of 'enoughness.' The number of 'bads,' as in low quality of production or increased pollution, multiply faster than the number of goods. Past an optimum scale, continued growth is stupid in short run and impossible in long run. There are limits on resources and certainly for impacts on finite, already-stressed ecological systems. We should consider reducing the number of goods.

We are at the point of enoughness, where frugality and discretion look attractive, where simplicity is appreciated and acquisition resembles a disease—that is the meaning of the word after all, dis-ease. People could forgo having too many things and cultivate voluntary simplicity, living in a way that is outwardly simple and inwardly rich, by consuming less. This attitude would require new values: Stewardship, not progress; austerity, not affluence; permanence, not profit; responsibilities, not rights; people, not professions; betterment, not biggerment; and enoughness, not moreness. Expressing these might make people happy.

People are often comfortable when they are attached to a place. Attachment to place is a form of deep love, from which many other virtues for living well spring, such as frugality and humility. Place allows us to participate in a symbiotic connection to the living earth. Participation means living with other species.

Why we Still Fear Wolves Or *Boo!*

For Europeans and Asians, the fear of wolves goes back in history for many centuries. Some fear and hatred exists today even as civilization has simplified and dominated much of the planet. Michael W. Fox notes that 'lupophobia' is not shared by indigenous Native American Indians or by a growing majority of citizens who oppose wolf hunting and trapping. Those people understand the place of wolves in nature.

On the other hand, there are thousands of applicants for state licenses to kill wolves, who see the wolf as a trophy animal taken for personal gratification, or as a pest to be 'sustainably harvested.' In a perversion of modern mythology, they praise the wolf as a 'worthy adversary,' although wolves have no access to helicopters or automatic weapons, whose function is to decorate cabin walls.

People who want to protect wolves in their habitats are mocked as tree-hugging Bambi-lovers that threaten their way of life and right to blast wolves. Killing wolves affirms their kind of manhood and survivalist skills.

Their ignorance about the balance of nature, wolf-deer and prey-predator relationships is heroic and self-maintained. Killing wolves to have more trophy deer is not a form of conservation.

The poor threatened wolf hunters now beg government to protect their rights. If we could only show them that their fate is linked with wolves and trees they might realize that wolves are a needed part of diversity.

Many deer hunters, already see themselves and wolves and other predators as essential components of healthy ecosystems. With such an ecological perspective they begin to articulate a hunting ethic, which acknowledges the vital importance of wolves, humans and other predators in helping prevent deer overpopulation and loss of biodiversity. This is especially germane considering that across much of the U.S. the white tailed deer population has risen over the past century from some 300,000 to an estimated 27+ million. There is a place for hunting, and all deer hunters are not Bambi eaters. The traditional notion of co-existence, is being promoted by organizations such as Project Coyote.

Grazing on public lands means sharing it with wolves. Protection is important but it can be done with cowboys and shepherds—not lethal poisons sprinkled everywhere or by blasting every seen predator.

Every diverse culture has sub-cultures and cults defined by demographics, economics, religious beliefs, education and values shared and opposing. Good governance accommodates such diversity to maximize the good of the nation state, including proper management of natural resources and public lands.

The government sanctions and funds the ranchers' war on wolves and other predators. It permits hunters and trappers to kill wolves for sport and fur pelts. This violates the public trust. The majority of people want wolves in the system.

Let's get educated. If we can not live with wolves in our house—the big house, wild nature on a wild planet—then we will never have a truly civil society with compassion and reason, justice and respect in all our relations and relationships.

Consider Economics in a Time of Serious Adjustment

If our population does shrink, how will we cope? Will things fall apart, as they have historically after famines or collapse)? Our business model requires perpetual growth, not only of population, but of goods for each buyer in the population. A growing population means a growing number of consumers and laborers. So, everything increases, but it is an economic pyramid scheme, dependent on perpetual growth and on the fantasy of a world without limits.

Can we have prosperity without growth? Yes, if we think about it. We worry that fewer people means fewer payments for social security, for instance. But, social security does not need to be tied to a growing youth or under class; it can be guaranteed by a change in the way government works. Government can guarantee employment and security by providing health and wealth. Security is also tied to the capacity for alternatives in work or retirement.

A declining population may put some strain on the built infrastructure, by requiring less of it. But, with the loss of some jobs, other people will become available and could be put to work. Large numbers will need to work on the transition to smaller numbers, dismantling highways or buildings. Large numbers will work restoring domestic or wild ecosystems.

Workers worry that corporations will cut jobs or lower salaries. But, the remaining laborers will become more valuable, and might campaign for higher wages and even fewer hours. Reducing full-time work to 28 hours per week, for instance, would also lead to full employment (statistics reveal that only about half a day's work is effective). Given the costs of recruiting and retraining, job security should increase.

Smaller cities, with smaller infrastructures, will fit the local ecosystems much better, allowing more of nature to provide 'free' services. Populations would also fit better. Fewer raw resources would be needed; recycling old materials would make more sense than finding and refining new ones. Less agricultural land would be needed for food and fiber, so the conversion would be slowed or reversed. Much agriculture could be moved into cities. Certain parts of the infrastructure will depend on the economies of scale, such as sewage treatment plants, although in that case artificial wetlands could replace the large artificial treatment systems. It might be easier to build arcologies while dismantling ill-placed or badly designed large cities.

Prosperity has not always depended on growth. Many earlier cultures were able to become prosperous with stable populations. The economies of Germany and Japan, from the 1940s to the 1990s, recovered and prospered while the populations were contracting. Russia experienced something similar, supported by new oil reserves and energy sales.

Financing prosperity may seem more problematic. But, the characteristics of populations may influence that; for instance, older people tend to save more than younger ones. Savings could help finance public works. Old people are no more of a burden than children; both can have separate infrastructures and be unemployed. We need to remind ourselves that prosperity is more related to the quality of life than on what can be bought.

Overpopulation and overproduction of carbon have been growing for a long time. Even if we could reduce them immediately, the momentum of growth would continue. Negative growth will not be as difficult as positive growth has been. Positive growth is just a habit. Negative requires less of everything. In negative growth, corporations would break up and be more diverse. With declining populations, natural capital could regenerate. Agriculture would again be linked to natural capital. Some cultures would have the opportunity to pursue traditional lifestyles. Other cultures could experiment with alternate forms of energy and ecological cities. If we decided it didn't work, then we could repopulate our minimal concrete termiteria.

Tame Banks and Corporations

Many people agree that too much money has accumulated in too few hands, like a boil on the nose. And those hands lost a lot of the money by gambling on bubbles and worthless packages of debt. Almost everyone lost money and acquired debt in the grandest modern pyramid scheme in the new century. And, now there is $50 trillion in debts. And, now everyone is afraid to change because this is the only financial strategy ever tried, that we can remember, although our cultural memory is very short and spotty. There are simple rules that could be tried. And, if we all realized that it is an emergency, the rules might be implemented immediately and be very popular.

First, the banks. Banks would no longer have the right to create money or debt. A bank creates money for a mortgage for instance. It loans it into existence based on the promise to repay. They get interest, but 80% of that new money gets loaned out again. More debt has been created. The bank only has to keep 20% deposits on hand in bank. So, one rule would be to make the bank keep 100% on hand. Another rule would be to return the power to create money to the government. Which would spend it on infrastructure, including salaries. Or lend it to local governments. As a result, the government would no longer have to borrow money. Then taxes would go down. Government would spend for people. A third rule would be to forbid banks from gambling with their money on mystery packages or bubble opportunities to make profits. Gambling is a risky enough profession for individuals.

Then, the corporations. By masquerading as individuals due special treatment, for the most part, corporations have evaded punishment for their crimes, which are personal, legal, social, environmental, and regional. A few simple rules could improve corporations and make them model institutions, and not human individuals: Anchor the corporation in one nation (or state); limit their charter to one public service, such as turnpikes or food preparation; put a fixed time period in charter, e.g. 15, 25 or 50 years (much like a radio license); eliminate special privileges; deny any possibility of making political contributions; ensure the responsibility of senior executives (close loopholes); require transparency, and limit name changes, mergers, and orphan divisions. Also, monitor employees for their safety; monitor environmental safety; limit rights to transfer from communities without warning, evaluation or penalty; and force them to abide by taxation, such that they cannot not pay. For multinationals, based in a home nation: Require adherence to international and national regulations; monitor adherence to fair trade, safe worker conditions, and fair pay.

The government needs to educate consumers on wise purchasing of necessities or intelligent wants. There are, of course, other things that can be done to return corporations to offering worthwhile services. But, that might require a change in awareness, first.

Although these rules are common sense and easy to implement, banks and corporations will most likely resist. After all, money and interest payments are now flowing to the top 1% of the wealthiest who control them. The wealthiest will want to fight any change in that, as well as any transition to rule-based government. But, we need to change now, it's an *Emergency*!

Why are Problems such Problems?

There are so many big problems, from pollution and economic inequity to drawdown of water and the loss of diversity of wild and domestic plants. Water is a daunting problem; there is too much of it in some places, but in most there is too little; droughts, serial or extended, have been the downfall of many older cultures, in virtually every continent. Yet, we continue to irrigate wastefully, to use fresh water wastefully, and to neglect to reuse it, forcing water cycles into abnormal patterns. Our wasteful ways seem to have too much momentum.

Others have recognized the momentum of short-term self-interest, self-deception, perceptual limits, overconsumption, polarization, and destabilization. And, they are all certainly right to do so—but these urgent problems are interconnected! The problems are interconnected, so we have to address everything simultaneously, because solving one problem will only offer a temporary, limited solution as other problems affect the one in the center of our focus. We are forced to recognize that problems, as well as relentless change and radical uncertainty, now confront many species as well as human civilizations. It is recognition that we do not have adequate understanding, knowledge or control to solve any problem once and for all. These long-standing problems result from long-standing challenges to human health and happiness: Violence, stupidity, greed, forgetfulness—not just personal but cultural amnesia—old diseases and new, lust for power, and the addiction to consume. Some of these problems are the result of detachment. Corporations contribute to detachment by producing items that encourage isolation or virtual worlds.

We tend to think of problems as unwanted 'side-effects' of the wanted main-effects, but all effects are equal, as R. B. Fuller noted, and must be addressed as equal. A problem (from the Greek words 'to throw forward,' which is what we tend to do with them) can be considered as a question proposed for solution. Most things identified as problems are embedded in a network. There is not one problem, there is not one solution. Problems could be considered also as challenges that we must respond to continuously, in the process of living, not as puzzles that have to be solved once for all time. A challenge is a calling into question or a demanding task (defined as 'a call to take part in'). It is about consciously choosing to see what can be done, rather than dismissing a conflict as terrible and unsolvable.

Addressing a problem often has to do with a power struggle, which becomes part of the problem. If problems are regarded as challenges that require a social response, then some conflict could be avoided. The problems of cultures, of natural ecosystems, and of modern, industrial, corporate, urban civilization, have been documented quite thoroughly. We have identified most of the problems in the problematique, from erosion, pests, and fertility loss, to population migration and diseases. But, we have addressed them separately, using technological innovations or political adjustments. We have not dealt with them in a whole pattern. We have not understood them as equal parts of complex large dynamic systems.

Sometimes we forget the historical context, that the decisions of our ancestors saddled us with losses, just as our losses will encumber our heirs with deforested landscapes on depleted soils, despoiled by exotic chemicals and hazardous wastes, in a network of impoverished habitats with an unstable climate, and of course, compounded by large intergenerational financial debts. The loss of nature and the wild, that is, the loss of habitats and species, may be devastating to all self-renewing systems, including human civilization. The remaining losses, such as the loss of place and the loss of design, are fundamentally human losses. These losses, especially of uniqueness and diversity, tend to flatten our image of the planet even more than economic connections.

Are Some Problems Insurmountable (Is Energy)?

Consider energy. Energy demand will continue to grow as long as the myth of growth holds sway; and this myth requires growth for prosperity and equalization. As long as well-being is associated with growth, the difference between development and raw growth will be ignored. These assumptions lead to problems with energy sources and growing demand.

Energy problems include generation. How can we generate energy cheaply, after the oil is exhausted? How can we supply energy for higher demands? Even if we could capture what we want, it has to be held or stored somehow (and it has to be fitted to the limits of the environment to accommodate it). Oil or gas could be burned immediately, but solar would be limited to sunlight or back up systems (batteries or dammed lakes). For air transport, the source has to be carried on a lightweight, aerodynamic vehicle.

The cheap cost of coal could work against green technologies. So, if the past is any indication, and people demand cheap energy, it looks like coal will replace oil, when it's low, and then nuclear power or solar (including wind and tide) may eventually replace coal. Nuclear power may be preferred for high intensity, large-scale power, while solar would work for decentralized systems. Of course, if people were educated to understand their long-term self-interests related to climate, they might consider paying higher costs, but only if the self-sacrifices were spread out more evenly, so that the poorest people did not have to shoulder the burden of costs. Cheap energy is the goal and the desire. It might become a major problem, since ecosystems cannot accommodate too much excess energy.

Our desires are going to influence or force our decisions. However, with our psychological and social factors considered, people are still going to demand low-priced, on-demand energy. And, it is likely that that demand is going to force acceptance of the least expensive option, coal, which is also the most abundant, dirtiest and carbon-heavy. We think we understand the health and economic issues of coal, but not the trade-offs with health or climate chaos. Increased waterpower is limited by environmental circumstances, and it tends to destroy the local environment, specially the lakes, rivers and streams. Increased solar or wind power has higher startup costs, and has major impacts on wildlife and the environmental. The cost of nuclear power, fission or fusion, has high start-up costs, as well as high initial carbon release (from concrete construction). Problems with nuclear reactions include long-term storage for waste products due to their danger. The threats to health are serious. The technology allows for tremendously powerful explosions; this is related to incompetence, accidents and terrorism. Nuclear material has limits also, like coal. Unless they were extended by breeder reactors, they could be exhausted in less than 200 years (or possibly a 100 years).

Land area under the sea is tremendous and could be used for storage. Burning trash would not generate enough energy for civilization, even though most of it is paper. Trash contains usable carbon; we could reuse it or bury it. Hydropumping might be a viable source of large-scale on-demand energy. It has advantages over batteries or compressed air.

Just the use of energy involves other problems in other systems, from pollution to carbon emissions. Even if we should control carbon emissions and limit them below 350 PPM—or 240 until climate stabilizes—we might not be able to avoid the next glacial cycle or extending the warming for 25,000 years. Climate change is a geological phenomenon. Carbon dioxide is a natural product in a global carbon cycle. It is necessary for the development of the biosphere. We can imagine that we can control things, especially since we have seemed to have contributed to climate change with industrial emissions, as well as to an extinction spasm from ecosystem conversions, but we may not be able to. The system may be too large and complex. The laws of physics, chemistry and ecology have to operate in any system built by industry, but embedded in and supported by wild processes.

Rethink Plastic: Why is Plastic Such a Problem?

Sailing in the North pacific subtropical gyre (one of 5 high pressure areas in the world), Charles Moore found a floating trash island that went on for thousands of miles (perhaps twice the size of Texas). Much of Moore's 'island' was made of plastic, from fully formed pieces down to small nurdles. The gyres cover 40% of the ocean or 25% of the entire planet, and all of them are attracting islands of trash.

Why didn't the plastic break down in to its component elements? Plastic is a petroleum-based mix of monomers shaped into polymers. Other chemicals are added for inflammability or suppleness. Plastic is replacing iron and glass as containers; it is lighter and more easily molded. Every year we produce 450 million kilograms of 'phthalates,' used to make plastic soft and pliable (known to be toxic to human reproduction systems). They can leach from packaging and coatings. In some food containers and plastic bottles, phthalates are found with a compound bisphenol (BPA). We produce 3 billion kilograms of BPA every year.

Why is there so much waste plastic? Because it is light weight and long-lasting, useful for containers and wrappings, and not much is recycled. Only 3-5 percent of plastics get recycled. Glass and iron are more easily recycled. PET and HDPE (numbers 1 and 2) can be recycled. Plastic retains pollutants and gives off deadly vapors. Products made from plastic recycling are limited to carpet and boards and jacket linings. Except for some incineration, every piece of plastic made still exists, because no life form can break it down into nutrients. Recycling also uses resources and energy and creates pollution. But it does reuse resources and it is wiser use.

Plastic does not biodegrade, it crumbles into smaller fragments. Plastic can decompose in seawater. And, it can contaminate marine life at the molecular level. Samples of seawater contain styrene monomers, dimmers and trimers (which seem to be carcinogenic in mice). Plastic is moving into the food chain. The danger is eating it or becoming entangled in it. There are minuscule pieces of plastic, called nurdles (lentil-size pellets of plastic in raw form), in the water. By weight, it can total 6 times more than plankton. The pollution seems invisible and ubiquitous. Nurdles and beads are easily mistaken for food, can be ingested, and can screw up genes. They disrupt the endocrine system, so that some male fish and gulls have female sex organs.

Like sand on a beach, the entire biosphere gets mixed with plastic particles. These particles change the properties of water and soil. Plastics pollution at this level and scale is almost completely unrecoverable for recycling or breakdown (from burning or solution).

Should we stop using plastics? Can we? Should we try to collect all waste plastic? What's the alternative? There is one based on cellulose that degrades in a possibly less dangerous way. Would that solve the problem? Would cellulose plastic be edible by zooplankton? Would it degrade completely, especially before it reached the ocean? Or, do we have to invent another one? Most plastic is used for storage items, which could be replaced by glass. Glass has higher costs for transportation, but is safer for food storage or preparation; it is reusable for generations, and it can be blown into beautiful shapes.

Figure. Plastic nurdles in seawater (Photo credit: MIT)

Stop Growing

Growth serves as a mechanism of evolutionary adaptation, by carrying out genetic instructions for organisms in the environment; growth is also conservative and stabilizing rather than innovative and reorganizing. Sometimes, growth can be problematic, when a physical body grows too much flesh or a population keeps growing past the carrying capacity. Other things, such as ideas and wants, have no natural size to exceed, but their growth can cause other kinds of problems at inappropriate scales or locations.

Population growth occurs when the birthrate exceeds the death rate, or when immigration from other populations exceeds emigration. Under many circumstances growth confers a survival advantage to a culture. At some point a population stabilizes. A population of individuals grows, or overgrows, like a population of cells in a body. And, individuals, like cells, receive signals from the environment that tell them when to die. Unlike cells, however, which rely on chemical signals, these other signals are in the form of food supply, climate, and predators. Some animals try to avoid such signals by moving; some plants try to avoid those signals by growing larger or producing chemical defenses. Human beings have used emigration and technology to overcome those signals. We have controlled epidemic and infant diseases. Babies are overproduced and protected; the least fit are not eliminated. Medical science has increased the number of old and maladapted dependents of society.

And, we have become very good at converting ecosystems to food systems and using technology to control our home environment. Agriculture, for instance, sparked an acceleration in population growth. Urban cultures responded to population growth by intensifying resource extraction (which still continues with the green revolution and genetic engineering), by relying on more technology and other strategies. By increasing the population, agriculture increased widespread hunger. Despite famines that caused hundreds of groups to wobble or collapse, the overall human population has kept increasing.

The 'cowboy' economics of Capitalism depends on growth for stability. It is oblivious to human suffering and to the potential of catastrophes, as well as to small catastrophes now. It is ignorant of the planetary scale and problems. Continued growth of the "free market" is amoral and pathological, benefiting the elite of authoritarian regimes as much as the oligarchs of democratic ones. It refuses to recognize, much less to pay, all of its costs, such as depletion, loss of security or extinction. The entire system perpetuates mass poverty and inequity, and justifies it by blaming individuals.

Exponential growth is said to be bad, and organic growth is said to be good. In fact, although organic growth is better, there is little difference during a world crisis; both reach asymptotes of suffering. The metaphor for the economy used to be a simple mechanical model for turning resources into products. It was assumed that to be successful, the economy had to grow and turn a profit continually. Unfortunately, the assumptions of the model were also simple and failed to consider human needs and natural cycles, causing great suffering and disruption. These assumptions resulted in overgrowth, with increases in complexities.

Growth does lead to intensity at some stages of development in the life of an individual, group or species, but growth can stop, and development can also lead to creative intensity. Thomas Jefferson had suggested not saddling the future generations with adverse situations. We do not seem to have the will or foresight to make decisions for the second generation, or the seventh. We need to correct the overbalance in births. Toughness and ingenuity might be required. In an organic system, growth contributes to stability, but it cannot continue beyond maturity, due to real physical and biological limits. A mature system switches from growth to development. And that is what we must create.

Think Catastrophe, Expect Catastrophe!

Some of these steps may seem revolutionary. Although revolutions have the connotations of violence and overthrow, they can be as quiet and regular, and unthreatening, as the turnover of an axle on a wagon or car. Thomas Jefferson suggested that little revolutions, every couple of decades, or every generation, could make the experiment fresh, as well as break up unproductive hierarchies of power. We could start these little revolutions with many small steps, as long as they did not contradict cultural norms.

The first revolution has to be to adopt a new attitude. Adopting a catastrophic psychology for the nation, to address current and imminent losses, is less revolutionary and a more appropriate response to larger scale catastrophes, such as the national loses of biological and cultural heritages. Poverty and inequity are growing problems in every nation. These are reasons to adopt an extreme attitude towards survival.

In the case of industrial nations, we have been embracing excess for many generations, so that we are crippled by stress and sickness. Perhaps all we need is a diet. The essence of a diet is to restore yourself to health, by restricting unhealthy consumption. As societies and cultures may also be guilty of this kind of behavior, so they need to put themselves on a diet. Archaic and agricultural nations have been strapped by historical inequities and unfair trading. Their challenge is to avoid simply repeating the same errors and consequences in the rush to acquire minimum standards and wealth. The solutions for all nations include trying new kinds of balance for self-reliance, paying attention to cultural and physical catastrophes, and striving for better equity. Because of the extent of our overuse and ecosystemic conversions, and their effects on natural sources, this situation is an ***emergency***.

The nature of an emergency requires everyone to drop their normal activities and normal behaviors and to respond to a catastrophe. The catastrophe is usually quite evident, a wall of fire or a massive surge of water that will destroy or has already destroyed homes and people, as well as insects and birds, plants and animals, and their habitats. But, we are finding that not all catastrophes are fast, human-scale or visible. The effects of those changes make us uneasy but not adrenaline-ready; the changes are reflected in starving children, hotter summers and stronger storms, failing food supplies, and collapsing infrastructures. We seem reluctant to give the causes of these catastrophes the status of real emergencies, partly because the catastrophes seem like natural events, such as a warming trend, and partly because they are related to our industrial habits, which provide us with necessities, as well as with luxuries.

We have to learn to recognize and respond to these slow catastrophes, these invisible catastrophes and these very large and long-term catastrophes. And, we have to do it now, before they crest and become overwhelming. We can do it. We have the evidence that things are taking a downward turn (the original meaning of the word catastrophe). We have acted on a large-scale before, in times of a world war. We were able to treat war as an emergency and to encourage or enforce remarkable changes, such as rationing or job-remolding. We were able to take these actions without destroying our citizens or our cultures.

Why would this experiment work? Because life has over three billion years experience with changing and adapting, because human life and cultures have over 50,0000 years of practical experience, and because humans are immensely adaptable—if they can adjust to poverty and suffering, they can adjust to a few good changes. Perhaps it is already too late—limits have been passed and the catastrophes cannot all be reversed. We do not know, and may never know, but we can still act as if we were wise, as if doing the right thing makes a difference. And, we will have worked together to help others, to improve things and to make good places. If we act ***now***, this month, this week, this day, this hour!

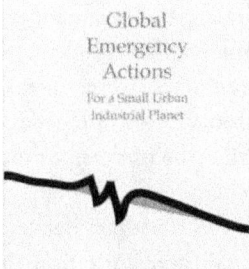

Figure. Book cover for Global Emergency Actions.

Avoid the First Awful Catastrophe

Our lust for growth and conversion of the planet is rapidly changing the social and environmental orders that represent the natural capital of social and environmental evolution. The reordering is also constructing an accidental trap that leads to massive catastrophes that could destroy that capital as well as the basis for its renewal.

We are already experiencing local catastrophes, such as earthquakes, fires and pollution, and a few regional catastrophes, such as tsunamis. And, we are aware that global catastrophes have happened in the past and are possible now. We know there will be a first catastrophe. We pray it won't be the worst.

We know that there are physical and ecological limits to the cycles and renewal of the planet. We know that we may have exceeded some of them. In complex, self-regulating systems very small changes have large consequences. In some cases, where conditions like drought are cyclic, in the Sahel region of Africa, humans expand during the good times, only to perish when the drought returns. In other cases, human activities, such as deforestation or overgrazing of herds, can cause weather changes. The large scale and slow rate of changes allows people to view the situation as natural, but once these catastrophes pass a threshold, the people and their cultures have been trapped by their demands, and only severe reduction or collapse can allow the system to regenerate.

We also know that not all catastrophes are short-term, single, small, fast, and visible. And, we are already in the midst of slow, long-term, large-scale, multi-pronged, invisible catastrophes. The climate is only the first that is becoming visible. Extinctions are another. Catastrophes can come in combinations, for instance, an asteroid strike, followed by volcanic eruptions and deep ocean releases of clathrates; possibly climate shifts, collapse of freshwater systems and droughts.

How will we act? Better than with climate or worse? What will people do when some disaster or austerity reveals their latent poverty? Will it be resentment and the rage of spoiled children? Will it be denial followed by self-destructive violence? Will humor and resilience surface?

Catastrophes concentrate attention on a landscape and its people, and that is one benefit on human affairs. Ideally, it should not take catastrophes to precipitate corrective measures. The present circumstances result from the decisions of our predecessors to have babies, fires, televisions, tractors, and status. The understanding of catastrophe may let us avoid it or at least ameliorate it.

If civilization collapses, the struggle back to a technological society will have greater limitations. Accessible minerals will have been scattered; the gene pool will have been greatly reduced. Then it may be too late. Our species may die. It is hard to imagine all life on earth dying. Even the worst of catastrophes would leave a simplified ecology of mosses and slugs. Weed and pioneer species would prosper. Habitats would be ruled by the natural laws of ecology again. Perhaps herbivorous animals would build up large populations again.

Fortunately, most catastrophes do not occur in completely destructive patterns. Limited starvation occurs before total starvation. There will be uncomfortable smog before acid rains destroy all crops. These things could signal the necessity of immediate corrective actions. Our attention needs to be global, since humanity has provoked a global crisis of local crises, typified by tremendous losses. We need to apply slow, long-term, large-scale, multipronged, invisible vision and thinking.

Still, based on a lengthy history and statistical probability, the first awful catastrophe will be a series of droughts, followed by famines and then economic unrest.

It Is an Emergency Now!

Buckminster Fuller suggested that the appropriate tools, such as eutopian design frameworks or advanced technologies, might help human worlds to work, given the right emergency. We have been encountering the right emergencies for well over a hundred years, and we need to refine and try these tools, now. The emergencies that he considered are the human responses to a variety of catastrophes, such as fires or earthquakes. But, we are finding that not all catastrophes are fast, human-scale or visible. The effects of those large, slow, invisible, long-term changes make us uneasy, but not really adrenaline-ready. The changes are reflected in starving children, hotter summers and stronger storms, failing food supplies, and collapsing infrastructures. We must learn to recognize and respond to these other catastrophes. And, we have to do it now, before they crest and become overwhelming. We can do it. We have the evidence that things are taking a downward turn (the original meaning of the word catastrophe). And, we have the tools for the appropriate responses.

When will we recognize the emergency? When it's just a little worse? A lot? When the change is more drastic? When fatalities are over a million a month? When? Is there an emergency? Certainly there is if you consider the extinction of species, a thousand-times higher than background, or the premature deaths of millions of people every years. Certainly if we consider our civilized debt-loads and how little food and fuel reserves we have, or if you look at the natural devastations from natural events and poorly planned wars. It is an emergency. As the catastrophes are large, the emergency responses have to be large, also.

This is an emergency, requiring large scale, multiple approaches, with new technologies, massive conservation efforts, and microenergy solutions (which require participation), but not using old, unconscious assumptions and design traditions. There is no time to wish politicians into patterns of good behavior. There is no time to beg corporations to stop cutting forests or poisoning lakes and ecosystems—as Garrett Hardin pointed out, conservation goes against their self-interests, which are focused on profits. The planetary emergency exists due to a series of slow, long, large, invisible catastrophes that are resulting from the normal wildness and uncertainty of planetary conditions, made worse by human modification of and interference with planetary cycles and diversity.

We need to act on a global scale. We have acted on a large-scale before, in times of world wars. We were able to treat war as an emergency and to encourage or enforce remarkable changes, such as rationing or redefining jobs. We were able to take these actions without destroying our citizens or our cultures (or most of the planet). Although we want to respond with a warlike approach, we have been fooled by the fact that we cannot see an enemy—and fooled by thinking there is an enemy. We have been misled by the slowness and subtlety of the penetration of our defenses. We have been betrayed by our own desire to continue our industrial dreaming at any cost. Some people have noticed changes and have been crying alarms, but they have not been loud enough or persuasive enough. Everybody needs to be awakened; everybody needs to participate, everybody needs to sacrifice and work towards peaceful solutions.

Human life and cultures have over 50,0000 years of practical experience adapting and making changes, and because humans are immensely adaptable—if they can adjust to poverty and suffering, they can adjust to a few good changes. Perhaps it is already too late—limits may have been passed and the catastrophes cannot all be reversed. We do not know, and may never know, but we can still act as if we were wise, as if doing the right thing makes a difference. And, we can work together to help others, to improve things and to make good places. If we act now, this month, this week, this day, then the changes might be more effective.

Figure. Planting Trees poster
(Credit: UN).

Is the Earth Too Complex to Understand?

The inventor James Lovelock regards the earth as a symbiosis of global dimensions, represented by the goddess Gaia. The Gaian ecosystem is a network of coupled smaller ecosystems connected by global patterns of water and air that adapt to each other through feedback loops that regulate physical and chemical environment—all based on great bacterial ecosystems mostly invisible to us. As a global ecosystem, the planet has special specific characteristics: The atmospheric composition is maintained by cycles pushed by living beings, at a relatively cool temperature; ordered complexity from dynamic change produces a diversity of forms. Environments selected for complexity, or ektropic order, produce entropy as part of the process.

Lovelock considers the first role of humans on Gaia was simply to recycle carbon and other elements. Now, he thinks that it might be to communicate for Gaia or perhaps to assist with the controls. Lovelock has expressed concern that humanity can impoverish the whole system by reducing the total variety. Although the planet seems to be very much self-regulating, changes in the atmosphere as a result of human transformations of ecosystems and wastes could trigger a new equilibrium that would be devastating to human civilization. We sometimes err to think that humans are superior because of our fascinating technology, but that creates new problems. The problems have much to due with global scale and global time lags compared to the two-year horizons of human business. Perhaps another problem is simply too many humans.

Perhaps the greatest problem is an aging earth, according to Lovelock, who states that there is only a small chance now of reversing solar atmospheric heating. He calls the earth elderly, although he states elsewhere says that she was young 65 million years ago. Actually, if she is old now, she was old then; That would be like going from 42 to 43 years old (middle-aged) in a human lifetime. At that time, she survived a global impact catastrophe. Gaia is 3 billion plus years old. The sun will continue to increase its output slowly for another 4-10 billion years. Of course, the planet will most likely live another billion or two as a self-regulated, living planet and easily another 4-5 billion years as a barren, rocky planet. The planet is not a body or a cell (as Lewis Thomas suggested it might resemble); it is a more loosely formed system containing various sizes of ecosystems. Tim Flannery suggests that this looseness may give the planet a substantially longer life, without the decay of aging, than smaller tighter organisms like mice or humans.

Lovelock states that soon Gaia will not be able to adjust to increased solar output. Michael Whitfield and Lovelock calculated in their model that in less than 100 million years the sun's heat will overwhelm the earth's regulation, and the atmosphere will move to a new hot state with a different biosphere. However, that's not the same as dying, even if the biosphere is not as comfortable for human purposes. Later, Lovelock states that any catastrophe that causes the Gaian regulation system to fail could lead to a hot dead earth; human actions could precipitate that. Heat-loving plants and bacteria may not have a critical mass of living things to regulate the environment. A critical mass of life implied, related to a critical area inhabited—Lovelock mentions 70-80 percent of the surface might be required.

Whitfield and Lovelock also point out that self-regulating systems tend to overshoot a goal and stay on the opposite side of the forcing. If too much heat comes from the sun, the system regulates on the cold side of the optimum. In the past the planet developed this way through a complex web of feedback. For humans to keep the planet cooler, for our comfortable civilization, we will have to manipulate what we perceive as controls or triggers. We might enter another Ice Age, which might be healthier for the planet, but might be equally disruptive to civilization.

Lovelock refers to the interglacial state as a fever. For life, a cooler earth may be a safer response to solar increase, but there is not a lot of evidence that it was more productive during Ice Ages, even having a greater land area with vegetation and fewer deserts. Lovelock argues that Gaia has greater control during glacial epochs, which has a lower CO_2 concentration in the atmosphere, which he interprets as indicating that the biosphere was healthier and more productive, because a cold ocean water is more biologically productive. He states this without noting that the oceans are relatively biological deserts now, and life on land may be more critical for cycles. Some data of the carbon composition in the deep ocean indicates that there was less organic carbon being fixed. Was the CO_2 too low for plant productivity, even with a larger land surface available near the equator? He notes that the rainforest is an adaptation to recycle water in a warmer environment. And, it is relatively fragile. It is important for carbon sequestration. Ocean deposition of carbon and dissolution of silica rocks are important.

Consider that the ice caps cool the atmosphere and lower the sea level. More land is exposed in the equatorial belt, which absorbs more heat. Do trees make the difference, creating more clouds? Are cool ocean currents less cool in glacial conditions? Do they bring up more sediments or less? Is there less sea life than before? Was CO_2 too low for more productivity? We do not know. It has been suggested that Gaia was very sick, since current large desert areas have low productivity, as do large areas of nutrient poor waters. This is a weak argument. Life does not have to be at a constant maximum to be healthy and vital. After all, the human organism has layers of dead skin, as well as organs, like the appendix, that do not seem to have current uses. Life lives on an inorganic substrate that is dynamically changing, due to the environment, including planetesimal collisions and living forms.

Natural selection at the individual level forces emergent properties, such as self-regulation, at all higher levels, including the planetary one. But, the cycles connect all scales. Furthermore, the biosphere as a whole is self-organizing. How could a planet have offspring? Send seeds to another dead or nonliving young planet? A dying tree makes more pine cones and seeds. Bacteria, viruses, and humans have escaped the planet. What is Gaia expected to do? Nothing? Ensure the survival of life? Maximize biomass or biodiversity? To live and be healthy? Is any living being required to have any other goals, other than living?

Evolution makes mistakes and eliminates many errors. Natural events destroy billions of living beings. Is Gaia cruel, therefore, like Kali or Nemesis? Or, is that a problem with personalization? Earlier, Lovelock suggested that metaphors were crude ways of knowing; later he emphasizes that they are needed, however, to comprehend the earth. Darwin had described evolution as wasteful, blundering and cruel. But, cruelty requires consciousness. And, we humans consciously drive plants and animals extinct in a cruel way. Gaia only filters. Gaia is not a cozy mother and cannot be propitiated by gestures like carbon trading or sustainable efforts. We are not separate from Gaia. The earth can be benign like ancient goddesses, but also ruthless. Possibly, we cannot understand complex, long-lived processes.

It is not likely that we will have to or need to save the earth; it can save itself, as it has done before. But, we may need to save the environments that we like as we know them. And this is what design can do. Gaia needs ecosystems on land and water for self-regulation. And, this is what global ecological design can ensure. Although the Gaia hypothesis renews the idea that the earth is a mother for us all, and reinforces our understanding of interconnectedness of biological processes, it falls short of demanding our responsibility and relies too much on our consciousness. Will we try to become responsible as one species?

Can Humanity Avoid Madness?

There are too many billions of us doing the same thing, over and over, without thought. That's just simple survival. Hundreds of millions are poor and starving, but until everyone is rich, we keep using the earth, even if it destroys our habitats. That's not greed, that's just getting some profit from destruction, so all can prosper, for a while anyway. We can master Nature, with just a little more effort, before the wild is killed. We thought we had mastered it in the past, but we destroyed much of it. But, learning from history is not necessary, because we have better technology, now.

If all that seems irrational, it probably is. Do we have any intuition about the fact that we are converting the flesh of the earth into human flesh? Maybe a few do. Do we continue extinguishing species and places, repeating the same errors over and over, because we have become detached? Oddly all this behavior is juvenile; we have not entered maturity, where we stop the repetition. Are our actions, as Paul Shepard thought, the result of our mutilation by a society unable to cope with the adaptive changes from agriculture and urbanization? At the same time, modern medicine recognizes the components of this 'maturity' as being psychopathologies, such as delusions of mastery over everything, resulting in injury to all. Because nature fails us, in our fantasies anyway. This leads to the insanities of nationalism, war, and destruction of the home ecosystem, which feedback into our fears and fantasies.

As nature is simplified, human society becomes more complex, until children learn to recognize automobiles instead of grasses or worms. Isn't that odd? Have we ever asked ourselves why we have to occupy every habitat on every continent, then surround ourselves with concrete? Does that lead to madness?

There is madness in science, madness in economics. Many of our less wholesome behaviors, such as rapid technological change or wars, can disrupt tradition and lead to misuse of the commons. Government or private mismanagement of common areas, can lead to economic madness. Our image of big science—the scientist as tragic hero, isolated in chaotic nature, but strong in his proud individuality, perhaps driven to research by hubris and madness—is a barrier to any new vision, especially a small vision.

Paul Shepard suggests that the entire Neolithic revolution has trapped us in behaviors that only end in madness. The feedback is inevitable.

Direction	of	Progress	——>		
Concentration	Intensification	Disease	Stress	Decline	Madness
Simplification	Instability	Famine	Drawdown	Destruction	Madness
Territory	Defense	Male domin.	Military	Take-over	Madness
Surplus	Specialization	Technology	Novelty	Stress	Madness
Distribution	Taxation	Inequality	Insecurity	Slavery	Madness
Human order	Abstraction	Isolation	Stress	Drift	Madness
Knowledge	Habit/tradition	Manipulation	Control	Laziness	Madness

Now we have to ask, what happens after madness? Do we die? Do we change and get better? Do we stay the same and destroy everything. How would we act if we were mad? Better than consumers? Can we analyze our way out of the trap? Or do we try a new approach to everything?

How I Would Destroy my Country (If I Were Mad)?

I used to say the Pledge of Allegiance every day in school; it was part of the day. And, I believed it. Then, in High School it was a monthly thing. In College, European Culture and then African culture became more important for ideas.

Somewhere along the way, I seemed to have departed from the main stream of thought. Then I realized that what was happening was an insidious kind of self-destruction, with normal happy people in collusion with the mad captains of the industrial machine, stoking the engines with everything delicate to keep the fires of greed and acquisition. Then, I realized that if I were as mad as many of the Wall Street robber barons, that would be how I would destroy the country. As a physicist I understand the part of entropy in relation to ektropy, the ordering principle that leads to complexity. But, entropy by itself would probably leave a planet of industrial forms, unmoving and unfinished.

Here is what I would do:

One. Industrialize everything. Not just steel mills but farming fields; not just automobiles, but cows and chickens. Not just houses and skyscrapers but people's needs and wants.

Two. Use up all the resources. Of course, this process is guaranteed to use up every single bit of copper, iron, gold and metals that can be collected. Since the planet has a lot of iron and aluminum, those are just harder to get and process with the energy available.

Three. Use more and more energy. Energy is another problem and it gets increased at every step, until the vast amounts start to overwhelm fragile ecosystems as well as the night. All the oil, then the gas and coal, regardless of the pollution, destruction and death. We can just drill and strip with our big machines and get all that concentrated carbon energy. Get it all. Then pretend wind and solar is going to help.

Four. Then, addict everyone to the products of the industrial process, new shirts and pants, new cars every few years, new homes, and new devices, new phones and new televisions, each coming out with more functions and gadgets and colors. These profits, some torn from unrich people in other countries would be plowed partially back into the process, but the rest would be riches for the captains.

Five. Just when everyone was well-hooked and spending too much— some people do want more than their share, and get it through dishonesty or clever theft, but when everyone can see that heroic wealth and consumption, almost everyone wants it and will overlook a little or a lot of dishonesty and theft, if they could benefit later themselves—I would loosen up the rules that kept fundamentally honest people honest, so they could grab as much as they could. This would be an institutionalized thing (recognize the puppet Reagan here?), perhaps the collusion of economics and politics—the same 'greedheads' would buy their way in.

Six. Mention that we should have been saving all the plants and animals that we now know we need. But, explain that science can capture their genes and we can save those and reintroduce them later. If the canaries in the mine are all dying, we can clone new ones, more hearty. And if the environment is too hot and all the ice has melted and all the cool animals are gone, we can reintroduce dinosaurs to hunt.

Seven. Offer complex fantasy worlds to replace the one that is being ruined. A nation of addicts will leap at the chance to find camaraderie and meaning in games. From our arm-chairs and beds we can observe the world burning and comfort ourselves that there was nothing anyone could do. It was just fate.

Unless someone could trace it to me. But, the same plan is working for the rest of the planet. Good night, sleep soundly in entropy.

Figure. Destruction begins (Credit: Daily Mail).

Pan Ecology Series 1983—

Create a New Image of the Planet

The system may need a new image to alter perceptions mired in an industrial worldview. This is the realm of cosmology, using a coherent image to explain and understand the planet. New technologies can actually contribute to making a new image for an ecological cosmology. The planet can be observed and photographed from space. Computers can create databases of ecosystems and create simulation models. Mathematics can make the workings of complexity visible. The independence of cultures and nations, as a result of new forms of agriculture and technology, can allow microcosmic cosmologies adapted to local environments.

Cosmology incorporates a complete set of ideas about the nature and composition of the universe used by a cultural system. This idea of the universe provides people with an orientation in their cosmos. Cosmology forms part of the ideological system of a culture. Cosmology is a collective image of the universe (Ezra Pound stated that the image is an emotional and intellectual complex in an instant of time). It includes beliefs about the origin, structure and destiny of the universe.

Archaic	*Industrial*	*Ecological*
Humans are same as others (only the masks differ)	Humans are different from others, masters of destiny.	Human one species among many in a community.
Humans can learn to do human things.	Humans can learn to do anything.	Humans linked in web of connections.
Humans are limited by human characteristics	Humans can change when they have to.	Limits constrain aspects of human life and change.
Humans are the same as always, but they can improve.	Everything can improve. Progress never ceases.	Human can fit in carrying capacity.

These characteristic cosmological statements result in specific behaviors. For instance, if the universe is a machine that implies that anything can be fixed, or if that is too difficult, then something else can be substituted. That means that nothing has a unique value. The choice of an image for a cosmology has serious consequences. The price mechanism of modern economics ignores the possibility that something may disappear; it assumes substitutability, that is, one form of capital can be substituted for another (and this leads to the disregard for any one form, such as land). The modern industrial image of nature as a resource has resulted in pollution, material shortages, and environmental degradation. A culture that degrades its eco-system risks its own extinction.

Many, but not all, archaic cultures are a form of fitness and limitation. Most archaic groups try for adaptation before domination. For instance, according to Gerardo Reichel-Dolmatoff, the goal of the Desana Indians is the cultural continuity of their society in its place in the rainforest. In Desana cosmology, technology was useful for cutting branches and catching fish. In industrial American cosmology, technology has no bounds and can do anything it wants, and should. For the Desana, fishing is best, although rituals are important; for Americans, the industrial system is superior to all others, which should convert to it.

Modern technological cosmology, beyond being another kind of order, more linear and abstract, is wrongly considered the evolutionary successor to traditional cosmologies, and is displacing them rapidly. Where human understanding is still underdeveloped, humanity cannot afford to suppress the diversity of thought necessary for adaptation to the diversity of environments, or to eliminate ecosystems and the societies adapted to them. The ecological, social, and political problems of today do not have simple, disciplinary solutions. The problems are cosmological and must be solved on that level. But a single cosmology cannot solve all problems in all places. A framework that protects local cosmologies as important functions, that is fit around them, not as a replacement, but as a means of preservation, understanding, and integration, is proposed.

Desana cosmos from the Columbia Amazon
(credit: G. Reichel-Dolmatoff)

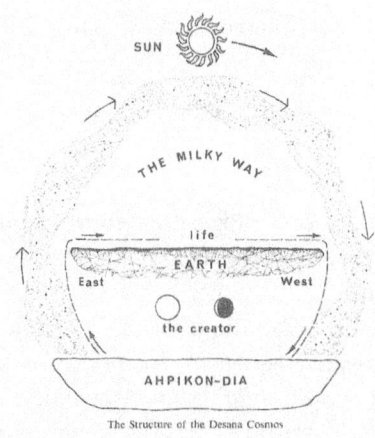

The Structure of the Desana Cosmos

How Much Wilderness Does the Planet Need?

Until we created artificial fields, cities and ecosystems, there was no need to distinguish between them and wilderness. We simply lived at home where we settled. Our impacts were not much different from other species who selected specific things to eat. Now that we have such a large population, with such dramatic impacts from our exploitation and energy use, we have become concerned with saving places where wild species determine the characteristics of ecosystems, especially since we rely on such systems for what we now refer to as 'ecosystem services,' the production of clean air and water, and the recycling of many wastes.

It is thought, if we save a large percentage of the planet in wilderness, that will be enough to automatically save both ecosystems and species. However, it will not save part of each kind of ecosystem. It is thought that we can save minimum viable areas of each kind of ecosystem, and that would save most of even the larger species. But, that will not save at least one population of each species. For that, we need to inventory all the species on the planet. Then we could calculate areas with Minimum Viable Populations (MVPs).

How Much Wilderness Does the Planet Need? That's hard to say, because there are so many human variables, not to mention evolutionary variables. The planet will continue, regardless of how we modify it. It may have fewer simpler ecosystems. What we are really concerned about, I think, is richness. Will we settle for desert ecosystems, blasted by centuries of war and destruction? Of course, these ecosystems have their own stark beauty, even after the adapted species have been destroyed. I think the moon is beautiful, but I would not want to live there for long, even with the best technical support.

The vast number of species existing now have existed for over a million years. These species have proved resilient to all kinds of natural changes and disturbances, except for human ones, which can interfere with renewal.

How Much Wilderness Can Represent an Ecosystem? I am not sure how much area an ecosystem needs to be self-renewing. Of course, that depends on the ecosystem. And, the larger and more complex the ecosystem, the greater the likelihood it has dependencies and exchanges with other ecosystems.

How Much Wilderness Does a Species Need? For species, we can calculate minimal areas to maintain minimal viable populations. But, even then, there are many variables to consider, depending on the species. For example, some species, such as wolves, use nonbreeding individuals to help with education of the young.

Michael Soule has suggested 500 breeding individuals as the Minimum Viable Population (MVP). The real size may be much larger, especially where the generations are shifted to elderly or where not all members mate.

Long-lived species are also vulnerable, for example, the Pacific Ridley sea turtle or African elephant. So, we might safely use 2500 as the optimum number for those populations, although that is far below the numbers that existed before human interference, and we are not sure how that is related to richness and stability. Also, we are not sure how big the minimum size of area is that will sustain an Minimum Viable Population.

It might be best to use a strategy that combines all three approaches: A percentage of the planet, in addition to special ecosystems and then special species. Without knowing how much wilderness the planet, ecosystems or species need, it may be possible to calculate an amount that might serve those needs. A rough calculation is 50-55 percent of the land area, and 75-95% of the ocean, including fragile shoreline areas.

How Much Wilderness Should We Save, Given Uncertainty?

Obviously, we should be cautious and save as much as we can of every habitat. If the area is large enough, then probably the top predators will survive and keep the system diverse and healthy. The current reserves are inadequate; wild species decline on limited reserves. Designing reserves requires a biogeometry, knowledge of the shapes of ranges. Left to themselves, in an adequate preserve, species may be dropped or generated. There are many ways to approach minimum (or optimum) wilderness areas. Some ways are simple, and others are complex, perhaps too complex. Some simple ways are probably unacceptable to the human populations. Several approaches can described as follows:
1. The Remainder method (*What's left?*)
2. The 50 percent rule (*Best guess?*)
3. The Acreage method (*After Eugene Odum*)
4. Minimum System Method (*Processes, after Lynn Margulis*)
5. Key Species: Top predator home range (*With genetics/ecoregions*)
6. The Mixed Method (*Appropriate for various areas*)
These approaches are expanded in the following paragraphs.

1. *The Remainder Method.* This method works by calculating how much wilderness is left and not critical for human needs. This would result in about 30-36 percent of the planet, according to satellite information. Early satellite information revealed that 32 percent of the land area was not being used in a way that compromised the systems self-renewal. A 1987 survey by the Sierra Club revealed that about 34% of the land area is undeveloped wilderness in blocks of 400,000 hectares. Over 200 international scientists identified 37 wilderness areas that represent 46 percent of the Earth's land surface, but are populated by just 2.4 percent of the world's population, excluding urban centers. To qualify as "wilderness," an area must have 70 percent or

more of its original vegetation intact, cover at least 10,000 square kilometers (3,861 square miles) and must have fewer than five people per square kilometer. Five of the wildernesses are also biodiversity hotspots, home to thousands of species found nowhere else. However, this does not tell us if it is enough to guarantee the functioning of representatives all ecosystems, for minimum function.

2. *The 50 percent rule.* The land area of the planet is over 149 million square kilometers (over 57 million square miles). Using the 50 percent rule, we should set aside 74.725 million square kilometers (over 28 million square miles). The oceans would also be protected at a much higher rate, probably 75%. Water area is over 360 million square kilometers. So, 270 million square kilometers would be preserved from extreme fishing, dumping or other use, especially those areas close to shore and upwelling zones.

3. *The Acreage Method.* Eugene Odum (1970) suggested using land area as a measure of human carrying capacity. Odum was one of the first to consider the implications of such limits. The minimum per capita acreage requirements, with a temperate area like Georgia as a model for a quality environment, is 2.02 ha (5 acres), divided into five areas: Food-producing (30%); Fiber-producing (20%); Natural support (40%); and, artificial areas (10%). Using this figure for wilderness of 40 percent per person, and considering the difference between mean productivity, we can calculate an optimum wilderness area for the planet, using the current population (as of 9/2003, Land area=149,450,000 km^2, or 56.77% of land area). This calculation results in a figure of just over half of the area of the planet for wilderness. If we use Odum's whole figure requirements, not just the natural areas, then the situation is revealed to be more critical; land requirement is then 131%. The only option in this case is to reduce the human populations over several generations—perhaps by lottery, or reverse incentives, or allowing cultural autonomy.

4. *The Minimum System Method* (Process). When the ecosystem is the smallest unit capable of recycling the elements of its membership, then for each type of ecosystem, a minimum area can be calculated that would permit the processes to continue to function. For example, organic carbon can be expired, fixed, reacted, or transformed. This method requires intimate knowledge of an ecosystem. For most ecosystems, we do not have that knowledge. Until we do, we should use a more conservative approach to determine a minimum or optimum.

5. *Key Species Method* (Top Predator and Genetics). By calculating the minimum viable area for each kind of ecosystem, using the largest habitat of the top predator, and summing for all ecosystems, a figure can be calculated for a wilderness area for the planet (if wilderness is equated with ecosystems MVAs). The idea of saving wilderness now is the idea of saving ecosystems complete with predators, most of which cannot compete with humans. Yet, the systems are crucial to human survival. The conclusion then is to save

large systems, as wilderness, complete with predators, not only to keep the biodiversity strong, but to allow those systems to provide those things that humans cannot do without, such as clean air and ecosystem services. The ecosystems have to have viable populations to be stable.

This system is quite a lot of work. But, the calculations can be simplified using three levels of tertiary predators across all identified ecosystems. For example, the US has 233 distinct ecosystems. Wilderness for the US, using three levels of tertiary predators for the 233 ecosystems, can be calculated at 5,243,500 km². The total area of the US is 9,373,000 km². By this calculation 55.9 percent of the country should be set aside as wilderness. By extension, that percentage could be extrapolated to the rest of the area of the planet (.559 x land area plus a large calculated sea area).

There is another way to calculate Minimum Viable Areas. That is to sum the requirements necessary for each species, since each species is genetically unique, not simply address the top predators. This is considerably more difficult, since we do not know how many species exist, where they are, or how much home range they need; nor do we always understand how the areas overlap. This calculation will have to wait on a complete inventory and monitoring program for the planet.

6. *The Mixed Method.* A mixed method might be more appropriate given the differentials in wilderness saved, unused but unprotected areas, and humanized landscapes. First, set aside those areas that are essentially wilderness already. Antarctica is especially important because we do not really know if it is possible to save only half of it, while drilling and utilizing the other half. Because it its influence on the water level of the rest of the planet, as well as political uniqueness as a research continent, it should be preserved intact. Other parts of the Arctic, Sahara desert, and Micronesia, should also be set aside, that is, be limited in terms of not increasing human impact.

Therefore, first we add: Antarctica; the Arctic parts of North America, Europe, and Asia; Amazonia, Central Africa, Asia; and Micronesia and Australia. Then we add hotspots of diversity: Madagascar; Klamath mountains, Ecuador, and others. Then we calculate areas of wilderness from the remaining areas, in terms of nations: United States, Russia, China, Europe, and others. Finally, we calculate the areas to be restored to minimum wilderness areas (in the categories, neopoetic and restoration, from the previous system): Europe, Asia, Africa, and others. This could be done in terms of ecoregions or ecosystems. As percentages, by Division, which allows finer distinctions, this results in 90,061,000 or 60.26 percent of the land area.

Why Save Wilderness at All?

Wilderness used to be blanks spaces on maps. Yet, as Aldo Leopold said, "to those devoid of imagination, a blank space on the map is a useless place, while to others, it is the most valuable part." These blank spaces, least influenced by human activity, are wilderness. They have their own functions and ultrahuman needs. We are learning that these blank spaces may be the most valuable. Wilderness has been appreciated for its intellectual, emotional, and economic values, especially as a source of materials and passive benefits, but not for its own sake. There are three fundamental reasons why wilderness should be left unoccupied.

First, wilderness is original and spontaneous. It is composed of systems that renew themselves and regulate the process so that structural integrity is maintained over time. The ontology of a living system is the history of maintenance of its identity through a continuous self-making process (autopoiesis). Living systems are self-organizing and self-reproductive. Adapted systems are optimally resistant to the forces that elicited self-organization.

Second, the basic unit of feeling in a system is the individual, a unique pattern of activity within a changing process. All living individuals develop images of their environments. Genetics provides the proper image choices for some, frogs, for example; others, such as coyotes, must learn what is valuable by using their senses. An individual has its own world that is strange and fascinating; von Uexkull called it an *Umwelt*. The world (life-image) is what has meaning for an individual organism. All organisms, regardless of complexity, are "fitted to their unique worlds with equal completeness." Simple organisms have simple worlds; complex organisms have well-articulated worlds. The human world is only one of countless actualities (and possibilities). Organisms are not suboptimal beings relegated by evolution to second-rate habitats; they are optimally fitted in places where humanity cannot fit.

Nature has been regarded as a pyramid of death, an arena of slaughter by biologists from Darwin to Birch. Nature has been regarded as inhumane, but it is not. Individuals engage in varieties of negative, positive, and neutral interactions. Cooperation is as effective and necessary as competition. Survival of the fitter is correct only to a point; beyond that it is survival of the more cooperative. Many organisms go to absurd lengths to avoid competition. Furthermore, most interactions are not simple but complex and paradoxical. Parasitism on an individual level may confer benefits on a species level; predator/prey are not excluding opposites, but generate a unity where there is stabilization and survival values for both.

Third, wilderness is the best stage for evolution, the process by which organisms and environments mutually adapt. Adaptation is not total; destabilization and risk accompany the process. Although evolution is seen as an ascending transformism, it more resembles a shuffling of individuals and environments. The net result is null; there is no transcendent spiral. Evolution goes in every direction. Plants and animals radiate through environments. Evolution is considered a building up of complexity, when it should be regarded as an unfolding of patterns. Evolution is not a hierarchical ladder or up-escalator, but a series of adapted forms. Evolution builds upwards and outwards, as well as inwards and downwards, from the simple to complex, as well as back again, creating incredible, nonreplicable diversity.

Empty space on earth is an illusion. Wilderness is full. But, wilderness is an abstraction; it is hard to define exactly or objectively. Any wilderness that can be described is also a state of mind, and because it is a state of mind, it is ambiguous. Wilderness is a changing process, and most languages become imprecise and mystical when dealing with process thought. Many attempts to define wilderness limit it to human experience, but wilderness by definition, is empty of most human experience. A true definition cannot be found in human needs and desires.

How Can We Protect Wilderness?

There are species that cannot coexist with humans. There are habitats that are critical to the functioning of the planet. There are habitats too fragile to bear human interference. There are habitats that are of virtually no use to us. All of these could be set aside and monitored from satellite (even airplanes can have some impact on some systems).

The wilderness that could tolerate small incursions, such as scientific research, would probably tolerate some human presence. Those wilderness areas that have coevolved with human cultures could simply be controlled by limiting human presence to traditional cultures or to sophisticated ecological cultures. Not every person in every culture wants a car and television. The allure of simple traditional cultures is such that many people would like to go back to them and live that way.

Some wilderness areas could probably tolerate a higher impact from people. Traditionally, borders (between human and nonhuman habitation) have tolerated human impacts. Most borders are between cities and unused areas. These are permeable and undefended, except for the few reserves. The boundaries are fuzzy anyway, due to the ability of pollution and trash to reach any area.

Even the idea of wilderness has been turned inside out, from the nature outside to the nature inside our scope, preserved from us and our activities. Confucius declared that fang-wai (outside the square) was the edge of his interest; now that we understand the interconnections, we need to reverse that attitude as well. What is 'outside' might hold the most interest.

The patterns we impose on all of nature now determine what species thrive and what do not, their numbers and ability to move and evolve. We tend to homogenize nature— homogenize is such an appropriate word—it means made the same, but also perhaps the process of sameness promoted by *homo sapiens*, the wise-ass homogenizer.

The numbers look great, but how important is shape? These areas have to enclose a large percentage of interior space to protect interior species. The areas also have to be lined up with watershed and airshed patterns. The context landscape is also an important consideration. Historical developments, such as drought patterns, should be considered; for example, to accommodate anticipated changes from planetary heating, reserve designs could have longer north-south axes (for a longer discussion of shape, see the book *Redesigning the Planet: Regions*).

We can Measure the World

We try to know the human place in nature through shape (anthropomorphism), centeredness (anthropocentrism) and measure (anthropometrism). Each way has advantages and limitations. The first human images of order (worlds) were wedded to shape (anthropomorphism) rather than to position, measure, or language. People saw human forms in every form of nature. The order of nature was a human order. Natural events, like lightning or rain, had needs and reasons. And these events could be controlled by human rituals, which satisfied the needs or influenced the reasons. Anthropomorphism gave human beings a place in nature. Other beings were seen as relatives. Anthropomorphism is the only way to understand mother, father and kin. Anthropomorphism leads to an understanding of the ultrahuman. We understand other beings by expanding ourselves, not by shrinking them. It is concerned with the projection of our experience into other beings, not with making other beings into us. Anthropomorphic thought increases the dimensions of the human intellect. Yet paradoxically anthropomorphism is limited. Humans project human need and thought patterns as guiding forces in nature. Nature, however, is not a human creation. Anthropomorphism is a personal interpretation of an order that encompasses all human orders.

An increase in knowledge from the neolithic to classical periods brought about a realization of the vastness and strangeness of nature. Concentration on human interests, anthropocentrism, resulted in greater success for the species. Humans became successful competitors. Then they became instruments of change in ecosystems. With Plato and Aristotle, nature became anthropocentric; it turned around a center: humanity.

Humans were the most important beings, at least through the Middle Ages. Then the Copernican revolution transformed the universe from geocentric to sun-centered and then centerless. The biological universe, however, was still a great chain of being where humanity was a link between beasts and angels. When Darwin linked humanity too closely with the beasts, cosmology became less meaningful. But the industrial, scientific revolution restored human importance by showing the power of human reason. Nature is seen exclusively as anthropocentric, as a human resource (especially by philosophers like John Dewey). Even the Biosphere Reserve Program is justified according to anthropocentric use. Much modern thought is concerned with inflating the position of humanity in the universe. Humans desire to make everything conform to images of themselves. All of nature is not human nature, however. There are many other sentient species.

Science is only beginning to support this idea. Plants and animals inhabit other worlds, other centers. The universe is not anthropomorphic, in the image of man. Nor is it anthropocentric, centered around man. But it is measured and valued by man, as it is measured and valued by all beings. Reference may be the rod that measures, but what is measured may be greater than humanity and its survival. Anthropomorphism is a necessary human way of knowing; all knowing is based on it. But the knowledge is not limited to just human experience. Anthropocentrism is the natural centering of human experience. But humans are not the only centers of experience. Anthropometric behavior is the statement that humans are the measure of all things. Not everything can be measured. But everything can be put together in a metaphorical language. All three concepts dealing with shape, center, and measure are needed for human knowledge. By rejecting anthropomorphism, the experience of others is restricted, and the scope of self-knowledge is reduced. Narrowness of experience is a source of human insecurity. By rejecting humanity as its own center, the experience of selfness is suppressed. By rejecting measure, perspective is lost. Humanity is embedded in an ecological world and needs these three comprehensive ways to understand it.

Figure. Measuring the World.

Humans are Omnivores—Eat with Conscience

Probably no consequence of human development has had a greater impact on the natural landscape and ecological processes than the production of food. Patterns of eating have influenced the constitution of species and the very contours of the earth. Throughout their history, humans have used animals and plants for food and clothing. Animals were followed, herded, corralled, tamed, and finally bred. Plants were domesticated later. As technologies developed, human relationships with animals and plants changed. Hunting, grazing, and agriculture provoked large ecological disturbances. Early domestic animals were revered, but nondomestic animals were considered competitors or nuisances. Now, animals are treated as commodities processed in factories and wildlife is regarded as useless. Hunting persists, but mainly as recreation. A few plants provide the bulk of human diet; the rest are considered ornamentals or weeds.

Although dietary habits were stable for long periods, they have been changing, for economic, personal, and social reasons. Many people limit their intake of animal products to milk, butter, and eggs. Some are vegetarians or vegans; others concentrate on fruits. Humans have been represented as omnivores, carnivores, or fructivores by different factions. In view of our control over animal and plant populations, a reexamination of our use of animals and plants is critical. There are advantages and disadvantages to strictly carnivorous or vegetarian approaches.

Carnivorism. Humanity has a long history of eating animals, and traditions of eating are important to the integrity of archaic cultures. Besides flesh, animals provide many high-quality materials that cannot be duplicated by an appropriate technology: wool, leather, lard, tallow, manure. Most animals eat food that humans cannot; they concentrate protein and convert low-quality protein to high-quality. Where animals are allowed free range, they graze in nonproductive places, such as steep, rocky slopes. In many places, wild animals are hunted; wild animals are adapted to range unsuitable for domestic animals and are usually twice as efficient in converting protein—an Oryx, for example, needs a third as much water as a steer and is immune to many of the diseases. Insects are an abundant, but neglected, source of food.

There are many arguments against carnivorism, especially as practiced in industrial cultures. Often, animals compete with humans for the same food, corn and soybeans, for instance; in 1980, livestock ate enough food for 14 billion humans. Intensive meat production causes tremendous organic waste and water and air pollution; moreover, factory farming methods are inhumane—calves, pigs, and poultry are squeezed into the smallest possible spaces. Currier Holman[1] said that his business at Iowa Beef Processors "is very much like waging war." And, as in war, the innocent suffer. The inhumane treatment of food animals results in lower-quality protein, while drugs and chemical additives, beyond the toxins and saturated fats already in meat, increase the danger to consumers. Circulatory and heart diseases are linked to a diet based on animal foods.

Vegetarianism. Limiting the diet to plants avoids the suffering associated with food animals. The practice is more efficient overall; more people can be fed per acre. By eliminating the cereals fed to animals, the acreage in production could be reduced by 51 percent. Grain is an efficient food, having a high calorie to waste ratio. Furthermore, technology could reduce the area needed to produce edible plants by 95 percent, with greenhouses, hydroponics, and alternate sources, such as algae. Vegetables offer other, untapped, sources of protein, for which appropriate technology currently exists: leaf protein, which is abundant, efficient, inexpensive, and suitable to tropical and subtropical growing areas; algae, which is efficient and protein rich; single-cell protein, from yeast and fungi, which is fast, efficient, waste-free and pest-free, and can be grown on petroleum waste. These new microbial foods could lessen the burden of land use. Furthermore, the use of wild vegetables could encourage the exploration for and use of neglected and unknown plants.

But there are disadvantages to vegetarianism. Plants are living organisms, also, and many plants are living when they are eaten, although sentience is a more important criterion. Many of the alternative sources, such as single-cell protein and algae, are deficient in amino acids or are of poor nutritive quality. Cereal crops, by themselves, do not provide a good balance of proteins, so many kinds of plants would have to be used. In some areas, this would mean importing protein, at the risk of economic imbalances and threats to local self-sufficiency. Plants also have poisons, to discourage invertebrates; still, many plants must be cooked before being consumed. Much of the land pressed into production by industrial agriculture is not suitable for domestic plants; habitats are ruined by the development of land for special crops, such as chocolate, tea, coffee, and tobacco.

Omnivorism. It is obvious that humans are not pure carnivores, but it is less obvious that they are not pure herbivores or fructivores. Physiological evidence, such as the shape of teeth or length of the intestine, points to an omnivorous existence for many thousands of years. Although some human cultures concentrate on large animals or on roots, most cultures depend on a combination of sources of food.

Vegetarianism is an ethical response to the suffering promoted by factory farming. But vegetarianism is not a compartment separate from industrial agriculture, social mores, cultural traditions, the rights of wild beings, and the necessity of sufficient wilderness. Diets are part of the cultural traditions that provide individual identities for all people. Cultures maintain regional differences and emphasize the unique social aspects of consumption; meals often provide important social and psychological benefits.

Many of the problems associated with human patterns of consumption are problems of scale, efficiency of exploitation, and a universal, commercial diet. Our lust for food has resulted in a war against other species, less reported than human conflicts, but waged more constantly, viciously, and mostly out of sight. We cannot eat without killing animals or plants. Human cultures are based on killing. Often, our wants and charities result in deaths. If we have zoos, to save a few species after destroying their habitats, then we must kill for those animals that are carnivores (as we do for our pets).

Knowledge of our carnivorous history should not paralyze us with guilt. Rather, as Raymond Durgnat says, "it's a reminder that what is inevitable may also be spiritually unendurable, that what is justifiable may be atrocious, that the best we can do will always be an organized butchery—and the possible best is itself light-years from fulfillment." Durgnat concludes that when we realize that society is "an organization of deaths as well as of lives, can we become more aware, gentle, alive, sad, free for a Schweitzerian reverence for life ... for all lives, and not just the next one."

Knowledge makes us aware of the costs of eating and, perhaps, inspires us to eat simply, as part of a simpler and more frugal pattern of life. To make choices, we need a way of calculating. Michael Fox presents a scale of suffering of animals: For example, dairy cattle are the least intensively raised while veal calves suffer the most inhumane conditions; turkeys have better conditions than battery hens. A conscientious person minimizes animal suffering by limiting diet.

Physical, mental, and spiritual well-being are dependent on a good diet. Being a vegetarian may be the best option for the urban residents of industrial cultures. For archaic or post-industrial cultures, where human needs are kept simple, limits are respected, and all beings are revered for themselves first, being a conscientious omnivore (the term is from M. W. Fox, who is a vegetarian himself) is a middle way that preserves the meaningful rituals of eating, yet uses animals and plants in an humane and optimal way.

Figure. Cattle & Erosion in Africa (Credit: M.W. Fox).

Can We Be Nonviolent & Defend the Earth?

Some groups of people, concerned with defending wilderness areas, eco-systems, and the earth as an organic body, have advocated using any means necessary. Earth First!, for instance, showed us that saving the earth requires more active participation than just letter-writing and circulating academic articles. They were among the first to put their bodies where their ideals were, and their stance has made some profiteers and environmental rapists more cautious.

Dave Foreman of Earth First! has said that monkeywrenching is non-violent resistance to the destruction of natural diversity and wilderness. It is not directed towards harming human beings or other forms of life. Many Earth First! tactics, however, make this statement questionable. Is Earth First! really nonviolent, according to Gandhi's definition and practice?

Mohandas K. Gandhi characterized his ethics of group struggle by the Sanskrit word *ahimsa*, meaning "nonhurting." Ahimsa is a closely related set of prescriptions and descriptions. Gandhi said it means avoiding injury to anything in thought, word, or deed. He adopted a wide interpretation of 'injury,' and included all living beings and nonliving things.

Gandhi mostly had living beings in mind, but injury to nature or to natural processes, could come under the general principle of *ahimsa*. Indeed, the concept of ahimsa is so wide that an act of violence to prevent other injurious acts, such as the exploitation of wilderness for profit, would be ahimsa. Yet, Gandhi referred specifically to destruction or sabotage as *himsa* (hurting), even if the things destroyed were not the property of anyone.

Positive action does not have to be bilateral or dualistic; it can transcend simple opposition and be positive without being adversarial. Gandhi was always willing to compromise on nonessentials. He character-ized himself as a man of compromise because he was never sure that he was right. Compromise is an essential part of the nonviolent person, satyagraha; *ahimsa,* as unselfish love, demands compromise. However, there are prin-ciples that admit no compromise, and furthermore, if the compromise fails, the *satyagraha* is ready for 'battle.'

Most of the thousands of direct actions on behalf of the environment have been nonviolent in the Gandhian sense and some have been effective. The "hug the trees" movement in India, for instance, physically blocked excessive logging in the Himalayas. The Chipko movement started out to preserve trees by embracing them before axes could be used and has resulted in a ten-year ban on tree-felling in over 550 square miles of the Uttarakhand in India, a major source of timber and water power.

Nonviolence is easily misunderstood. If your opponent respects violence in defense of property, and you misjudge your opponent and offer nonviolence, which is perceived as weakness, prompting him to violence, what should you do? Stay in the center of the issue and be active, but do not respond violently yourself. Most such opponents eventually recognize perseverance as an indicator of strength. Furthermore, compromise can be such that it satisfies the opponent's ego without giving up much, because it creates a state of cognitive dissonance, where a little token is enough to convince people that something is owed in return. It is important to formulate one very clear, concrete, easily understandable goal for an action and alert your opposite to that goal as soon as possible.

The code of nonviolence, as presented by Gandhi, is not a rigid system. Exceptions are possible and desirable under some situations. Arne Naess suggests that a small piece of a technical installation, a dam for instance, could be destroyed in order to avoid the greater destruction of an area. Nevertheless, this violence is an exception and not a norm. Earth First!, by compromising on nonessential issues and using violence only as a warning, might increase its effectiveness. Either way, Earth First! already exemplifies an important, and neglected, aspect of Gandhi's philosophy: *you should follow your inner voice whatever the consequences.*

Nature is Our Self

Many ideas of nature are not objective or scientific. They underlie thought. Their insights are embedded in language. For example, the root word of both ecology and economy is based on the Greek word for house; ecology is its study, and economics is its management. The word house is used as a metaphor for nature. Similarly, other metaphors are used for nature, so that nature is seen as mother, father, sister, brother, and self.

Some have suggested Nature as Mother and Father. Gary Snyder discerns an undercurrent in civilization since the late Paleolithic. He considers Buddhist Tantrism to be its finest and most modem statement: "that Mankind's mother is Nature and Nature should be tenderly respected; that man's life and destiny is growth and enlightenment in self-disciplined freedom; that the divine has been made flesh and that flesh is divine; that we not only should but do love one another ... these values seem almost biologically essential to the survival of humanity."

Homer sang "of Gaia, universal mother,/firmly founded, the oldest of divinities." The idea of nature as mother forms the basis of a modem, scientific hypothesis, to explain how the planet exerts a living control of the atmospheric and hydrologic processes to maintain minimum conditions for life over long periods of time. James Lovelock believes that there is a collective global mind (that he calls Gaia, however unconscious) immanent in the cybernetic structure of the global system.

Others regarded Nature as Sister or Brother. Saint Francis of Assisi, in

"The Canticle of Brother Sun," addressed the Sun, Air, Fire, Wind, and Water as his brothers, then the moon and stars as his sisters; he praised the earth as his mother. American Indians, such as Black Elk and Seattle, also referred to animals as brothers and sisters. All animals, "Two-legged and four-legged," were equals. The phrase, "all our relatives," was used in prayers and rituals referring to plants and animals as well as to human kin.

Is Nature Our Self? Our bodies contain the ashes of stars; human cell structure is shared with trees; we share our bodies with bacteria, fungus, insects, many of which are beneficial--and even those not considered beneficial may have positive effects on our health. As Lewis Thomas shows, our human bodies are living communities, hosting amoeba in the blood, mitochondria in the cells, bacteria in the intestines. We are connected to the largest and smallest beings.

We have mistakenly concluded that our skin is the boundary to our selves. But, our intuition senses our interdependence with nature. We extend the boundaries of personality to other things and people. We participate in relationships in a field of relationships. Because we are in the field, the study of nature is, to some extent, the study of ourselves and our effects on the field. The organism is a point at which the field is focused.

We depend completely on the natural environment, physically and psychologically. D.O. Hebb has conducted experiments that show the effects of a limited environment. Cut off from external stimuli, the mind becomes strange and distorted. Mental health can be related to the quality of the landscape, as Rene Dubos and others (e.g., John Passmore, Paul Shepard, and Ian McHarg) have done. The external world is needed to keep us alive and sane. This world is composed of remote occurrences, on polar icecaps and distant stars, as well as immediate personal events. The individual is woven into the world.

Or Is Nature Itself? The earth has innumerable modes of being that are not human modes. Our direct intuitions of nature tell us that the earth is infinitely strange; it is alien, even when gentle and beautiful. It seems often mysteriously impersonal, unconscious, immoral, hostile, and awesome. J.B.S. Haldane recognized the strangeness of nature. "I have no doubt that in reality the future will be vastly more surprising than anything I can imagine. Now my own suspicion is that the universe is not only queerer than we suppose, but queerer than we can suppose." Perhaps the queerness results from sheer complexity. In its immense complexity, nature seems wholly other, nonhuman, ultrahuman. It seems distant. So it is feared as unfathomable and uncontrollable. Nature seems contradictory and sinister, shaped by death, which we fear.

We cannot approach nature or her beings as they are through our personal and economic interests, but only on their own terms, in relation, through respect and love. Any other approach separates us from other beings and truncates our aesthetic responses with boundaries. We are part of the cycle, woven into a poetic, mythic unity. But, the unity may not be comfortable, and nature is not a father, mother, self, or any entity of our wishing. It merely is.

Myths also limit human cultures, so that other beings (brothers and sisters) can make their homes in their places (father and mother lands) within the body of the earth. Wisdom cannot depend on perfect knowledge of other beings or places; such knowledge does not exist. Humans must act "as if" in Hans Vaihinger's words they were wise, that is, circumspectly, with caution and respect, as if nature was our very self.

Promote Ecophilia: The Metaphor of Home

Ecosystems are embedded in places. The making of places by human beings is an ordering of a distinct structure and center. The organization of perception, meaning, and thought is intimately related to specific places. Place is a focus of meaningful events and a platform for ordering a world. The individual image of a place is modified by memory, experience, emotion, imagination, and intention. Attachment to place is a form of deep love, from which many other virtues for living well, such as frugality and humility, spring.

A place is a part of the environment claimed by feeling. Humans, like plants and animals, identify greatly with local environments. Maybe this is a function of the limbic system of the brain, a function we share with territorial mammals. Human emotion creates an 'in-place.' Emotion binds together motion and perception. Emotion can transcend distance. A place that is found and made; it does not exist before.

Modern populations are rootless, moving about from city to city. We are suffering from a placelessness, which arises from our style of efficiency and proclamation of mass values. So far, no psychologists have studied what happens when a person sees her/his place, their very context, destroyed. These catastrophes may be the basis for diseases, depression, or cancers. The word nostalgia was coined by Johannes Hofer to describe an illness characterized by insomnia, palpitations, stupor, fever, and persistent thought of home. The disease could result in death. For some people, the Northern Aranda in Australia for instance, it is not possible to stay away from home indefinitely and still live. Thus far, the sense of place cannot be gleaned from an analysis of the nervous system. Yet a place shapes the nervous system.

Human maturity is linked to the increase of identification with, and care for, others and for place. Humans love place (topophilia), as well as life (biophilia) and home (ecophilia); perhaps this love is an instinct or a meme. The inexhaustibility of a living being or a living place constitutes much of the nature of love. Human beings are compelled to seek other beings and love is the most rewarding approach.

Paul Shepard suggests that for each individual the organization of thinking and meaning is intimately related to specific places. The place is a matrix for ordering experience. The specificity of place is important. Animals and humans are imprinted early in life to particular places. Every place has a unique identity. *Topophilia*, love of place (coined by Yi-Fu Tuan), is the recognition that all human beings have affective ties with the environment.

Intuition also senses the interdependence of nature. We extend the boundaries of personality to other things and people. The beauty and complexity of nature are continuous with ourselves. We know subjectively that we are not separate from the earth, that wolves are capable of love and tenderness, that trees are beautiful. Edward Wilson argues that the essence of humanity is inextricable tied to life on the planet. *Biophilia* (his word) is the natural affinity for life, and is central to the evolution of the mind.

We can coin a term to describe the love for living places, *Ecophilia*. One ecological benefit of rootedness is that people will take care of a place if they realize they are going to be there for a thousand years. Having a place means that the inhabitant has stock in it and participates in its unfolding, through planting and caring. Detailed understanding of plants in a locale allow gathering of food and medicine. People cultivating a sense of place are people in place. Their work can be appropriate; appropriate growing, logging, mining, or building. People in place acquire a sense of community, nonhuman and human; a shared set of values and concerns; health and spiritual benefit. They feel at home. When we see land as a community to which we belong, we will use it with love and respect.

Advertise Ecology as a Way of Life

Advertising creates the mythic images of our industrial cosmology. The myths are powerful, but trivial, they are memorable, but inadequate to convey the meaning people need to live. Perhaps the myths are restricted by their content, as well as by weak poetry and art. Few sciences and conservation movements have strong images broadcast regularly to billions of people. Ecologists and artists, as well as urban planners, historians, and politicians, need to use the strengths of advertising to convey ecological sense and traditional wisdom, the feelings of balance in a new mythology.

Business has transformed much of art and poetry into advertising, to match the style and attention span of the people in industrial cultures. Advertising, quite literally from the *Wall Street Journal* to college textbooks, refers to its activities as "shaping the American dream." Like art, advertising creates an image of a way of experiencing. Unlike art, it limits its focus for a specific goal—profit. Like art, it mirrors us. Unlike art, it intensifies and glorifies only the positive aspects of culture, ignoring the dark, negative aspects. And, like art, advertising lies (although Jules Henry thought it was instead a new kind of truth—"pecuniary pseudo-truth"—not intended to be believed or proved) in the service of profit-centered corporations.

Its simplicity is irresistible. Our environment deteriorates according to ecologists, but gets better according to economists. And their pictures are prettier. People want to hear that it is getting better. Advertising tells them it is. People want to act stupid, greedy, and selfish, and spend the inheritance of their children on themselves. Advertising tells them their actions are rewarded. The real issues of life and death, destruction and hope, make people feel helpless and anxious, so advertising points their consciousness to comfortable trivia.

Despite the ugliness of the dreams of progress and growth, of waste and stylistic frenzy, advertising, using sophisticated techniques and narrowing the focus out of context, makes the dreams desirable and irresistible. People in agricultural and hunting cultures interiorize the abstract industrial vision. African farmers are convinced to buy inorganic fertilizers, even though it degrades the soil; women to buy powdered milk for their children, even if it kills them. Tractors replace draft animals in the paddies in the Philippines, even though they are costly and less energy-efficient; French winter fashions are found desirable in tropical Brazil, even if they can only be worn in air-conditioned villas. People in industrial societies are convinced that their children will be ruined without personal computers. Disposability is offered as a fix to a wanting in the temperament. Advertising fuels the acceleration of conspicuous, compulsive consumption.

Yet, advertising may be the most effective means to reshape desires and reform buying habits. Advertising presents the symbols of modern experience, even if they are just the trivial ones. It could present healthy symbols equally well. Advertising does incorporate traditional values, like family, friendship, and love, although to sell beer and cereal and, sometimes, churches and hospitals.

Advertising is beginning to support more informational functions, such as the dangers of drug abuse and smoking. Advertising creates values—fur coats, fast cars, dark beer, slim cigarettes are certainly recent and artificial values—but it could be used to create positive ecological values and new identities that show that our needs for prestige, esteem, and belonging can be met without stylistic waste at mindless speeds. Advertising could promote new attitudes about appropriate technology, the rights of other cultures, and the place of people in nature. Good advertising could be as subversive and conservative as ecology. It could avoid confrontation with people's values; emphasize positive aspects without negative ones. A good ad could capture and carry the most self-indulgent viewer; for the most part, ads don't require effort, literacy, or consciousness—just attention.

To work towards this service, conservation groups could define and promote an integrative mythology as the basis for the framework of diverse efforts to protect life and the environment. Conservation groups could provide a meaningful philosophical foundation, as well as coordination for other humane, social, and conservation programs. But, the approach must be egalitarian: Respect for animal or plant life cannot neglect human life and suffering. Advertising could provide reeducation through the most effective means. Conservation groups could spend money advertising "humane consciousness," moderation, and the joy of living (instead of just consuming or winning). Ecological ads would be unique and compelling, simple and effective. They would advertise not a product, but a way; not for a profit, but for a dream.

Make It Sexy

Would advertising for ecological consciousness really work? The other day, tired of writing, I went to visit friends, who were watching a car race. It occurred to me that only with entertainment industries is there so much technical fireworks and coordinated enthusiastic teamwork. Imagine all that energy and enthusiasm directed to appropriate technology for reforestation or the proper use of forests. Imagine television coverage of forest work with the same amount of attention and detail. Why not a competition for the most beautiful or productive forest or teams working to restore devastated city areas—broadcast by a major network as an important event.

It also occurred to me that this remorseless entertainment is an anesthetic against fear, emptiness, self-searching, or death. Continuous entertainment is a kind of guarantee of health, riches, and long-life. Everything that is pleasurable, thought George Orwell, seems to be an attempt to destroy consciousness. Television seems intent on proving him correct. Ecology cannot ever compete with entertainment if it raises troubling questions or difficult expectations. As long as the industry can guarantee many wetlands and forests through the arithmetic of fantasy, we will always seem to be complainers and false prophets—until it's too late, then we will be blamed for not avoiding the catastrophes.

Maybe the situation is not that bad. Maybe we can present images that rival the industry images. Maybe we too can speak the languages of euphemism that large corporations use to conduct their businesses of larceny and fraud. Positive images and pleasing language skills are everything these days; no one really looks for substance. The devotion to money, beauty and youth is our focus. But, as long as old forests and smelly wetlands have limited monetary value, few will focus on them, unless we can subvert the devotion with facts and understanding.

So, we should use the means, but tell the truth. Not make it perfect and clean, but raw and wild. That should be sexy enough. If it's not, then we all have much more to learn.

Is Environmentalism or Ecology a White, Elite Plot?

As awareness of ecology has grown, so has awareness of its limits as a science and a movement. Nathan Hare characterizes ecology as an elite, "white" science. William Tucker echoes that environmentalists are elitists who preserve rich resources for their exclusive recreation. Both critics level all groups to the lowest denominator. But, these groups also can be discerned as part of a large ecological movement, with a unique philosophy, science, economics, and politics, and with a higher denominator.

Hare laments that blacks and their environmental interests have been ignored by the ecology movement; indeed, he states that the two stand in contradiction. Ecology ignores the needs of the poor. But, racial justice ignores nature as the basis of all wealth. Both groups are ignorant at great risk. Yet, both groups are largely ignored or underfunded by government. The wars on pollution and on poverty are as ineffective as the wars on people. In one sense, he's right; ecology and blacks often end up competing for federal crumbs from war machinery loaf.

Ecology seems to be the faddish successor to middle-class concerns for conservation, a distraction from poverty and war. But, ecological action and wilderness protection, like earlier forms of social action, such as labor laws, women's rights, and minority rights, have been the result of leisure-class elitism, the projects of those with enough to eat and time to think.

People in the "culture of poverty" (Oscar Lewis's term) have less concern for the future. Those who worry about being assaulted in their homes usually do not display as much concern for threatened species. Those who worry about their debts, jobs, or health are less likely to be concerned with acid rain or Amazonian deforestation. Worse, many responses to the environmental crisis have been superficial, dumping garbage into the poor neighborhoods or sending pollution out of state. Many environmental groups have been concerned with the health and recreation of their members or the maintenance of their own hierarchies of power. Reactions appear in minorities as an anti-ecology sentiment.

That sentiment is unfortunate. If ecology is unpopular it is because, at a time when advertisers are expanding our desires for things and pleasures, ecology is describing the limits of nature and the limits of humanity in nature. Industrial consumer propaganda is infecting everyone with the unfulfillable desire for temporary goods and flashy mobility. The false hopes of industrial culture do more damage than the uncertain warnings of ecologists. The problems of pollution, overpopulation, environmental degradation, and inequities are the result of disregarding limits while increasing our domination and control of nature. Our successes in medicine, agriculture, and technology are unsustainable.

Only in a dangerously unbalanced economy is well-being for the poor tied to more wealth for everyone, especially the wealthy, and is adequate employment tied to expansive growth. The purpose of capitalism is acquisition, at any cost, even at the destruction of a rich and unique heritage, and not equal goods for minorities. Industrial growth threatens the fundamental structure of the environment that supports humanity.

Hare states that blacks suffer when colonizers use the resources and labor of the colonized to develop and improve their own habitat, while leaving that of the colonized undeveloped. This is certainly true, but the discrimination occurs at a cultural class level all over the earth. The underlying institutional structure is the enemy, that structure common to the war on poverty, the war on weeds, and the war on other cultures. The structure of industrial institutions threatens everything with an expediency that has been used to justify slavery as well as wilderness destruction. The prisoners of addiction (rich) and the prisoners of envy (poor) are both slaves (Ivan Illich's term) in a consumer society.

Racial or class groups that seek a share of the unbalanced economy perpetuate the inequality and hierarchy of the industrial state and legitimize the institutions that make the pollution and thrive on slavery. But, the economy cannot be divorced from ecology; the myths of limitless growth and free goods cannot continue. Ecology seeks to conserve goods, and the greatest of those goods is a self-renewing and self-balancing nature. Ecology is more than a pollution warning for restricted beaches or special devices for internal combustion engines. It is a warning about the imbalance of ecosystems and global cycles.

Many of the things that humanity has to face, in terms of limits and discipline—population control and austere consumption—will label the environmental movement "anti-poor" or "anti-people." But, that doesn't negate the importance of the movement or mean that it doesn't offer the best chance to avoid greater misery and greater catastrophe. The effects of this movement are more immediate to the poor, since they have less insulation against pollution and catastrophe. It is true, also, that many solutions to improve economic quality will have more adverse effects on the poor classes than on the rich. Ecological economics cannot be considered without some notion of distributive justice.

The warning to alter the trends of consumption and usurpation cannot be used as a rationalization for continuing the present social inequities between blacks and whites or between classes or between hemispheres. Hare and others have complained that the environmental movement is a "cop-out" in the struggle for justice. Hare is right in stating that social justice must be established before any solution to environmental degradation can be found. The human victims of greed, violence, and oppression, need justice. But, long lasting justice is impossible without the proper ecological relationships and responsibilities. Without ecology, social changes are doomed to be short-lived and painful. Nor can ecology ignore social injustice. Without social justice, ecology is doomed to be impotent theory. There are two houses for humanity, the urban system and the life support system of nature. Ecology attempts to recouple both into a harmonious whole. A rat-infested tenement is part of urban ecology, as much as spotted owls are part of old-growth timber.

Ecological justice cannot be created without economic changes, in pricing and profits, as well as in the counting and discounting of real resources. Furthermore, reforms in justice advanced to manage the crisis, rather than eliminating it, are cosmetic correctives to an irrational, unbalanced society.

Ecology can be a revolutionary process, based on the need to reconstruct industrial societies within ecological limits. It is not landscape decoration, as Hare dismisses it; the ecology movement is trans-class, embracing all economies and ideologies.

The concern of ecology is not elitism as Tucker charges, but the growth of institutions, energy use, population, consumption, and waste, resulting in the theft of tradition, uniqueness, choice, a convivial environment, and wildness. Growth, greed, and apathy are social and ecological issues, not simply class or race issues. Economic inequities and class problems are the more visible symptoms of social and ecological imbalances. Freedom, empathy, and balance with nature are not specific to a class or race, or gender or species—these are universal interests shared by all beings, human as well as ultrahuman.

More than any other movement, the ecological movement needs to mobilize popular involvement for environmental health and balance. Humanity will not die without gadgets, jets, or televisions. Humanity may not die without clean air and water. Humanity *will die* without meaningful work or play, or without meaningful relationships with others and with wild nature.

Figure. The plot thickens
(Credit: Forbes Magazine).

Promote Radical Ecology

The concern of radical ecology is to ensure the survival of human communities in place on earth, which is, in fact, the goal of politics. Because survival is in nature, politics must rest on an ecological foundation. As a science, ecology describes the interrelationships of organisms and environments, that is, the experience of living together in the biosphere. As a philosophy, radical ecology investigates the normative aspects of living together, that is, ethics, and the maintenance of the affairs of a community, that is, economics and politics. As a noetic discipline, it provides information on the state of nature, but recognizes that human beings are participants in nature, as part of the food chain, for example, as well as participants in the societies that are trying to survive. It offers a new perspective of humanity in the total field of nature and defines balanced relationships with ultrahuman beings and species. Radical ecology addresses the determination of separate wilderness areas necessary for a healthy ecosphere, and an optimum human population, based on net ecosystem productivities and modified by appropriate technologies within ecological and cultural restraints. It urges local, self-reliant cultures with adaptive cosmologies and natural values.

Modern science has provided us with the means to manipulate natural processes for human advantage. But we distance ourselves from what is uncontrolled or unowned. This detachment is the greatest threat to the welfare of nature and, ultimately, ourselves. The vivisection of the world depletes our ability to feel compassion for it. We are "destroying the voices of existence," says Neil Evernden, without even hearing most of them.

Ecology deals with the relationships of organisms to environments. It is not a reductive discipline, and not readily amenable to quantification. Even scientific ecology is an integrative discipline that extends beyond the bounds of science. Ecology is an amphibious discipline, with the authority of science and the force of moral knowledge. Ecology, studied through its components and relations, is a perspective, a way of "seeing," according to Paul Shepard. It is a perspective of the human situation in its interconnection. For Paul Sears, ecology is a "subversive subject." It is normative and sensible. But for Theodore Roszak, it offers a "sacramental vision" of nature.

Radical—from the Latin word meaning "rooted"—ecology forms part of a new metaphor that is more appropriate to the unity and interrelatedness of the earth. It is part of a movement of consciousness, concerned with equality, diversity, health, with humane methods, and with a holopoetic cosmology. And it affects them simultaneously, as Henryk Skolimowski recognized of eco-philosophy.

Radical ecology emphasizes biological equality, like the "deep ecology" of Arne Naess. Charles Elton transformed the Great Chain of Being into a chain of eating. The result of Elton's food chain was the realization that the bottom link—plants—is the most important. Humanity is part of the chain. A critical message of ecology is that if we diminish variety in the natural earth, we debase its wholeness and stability.

Radical ecology incorporates a broader scientific method that might be called patient practice. There are ways of dealing with the earth that are not scientific or technological; they are aesthetic or ethical. These alternatives are not incompatible with a whole science. The methodology of traditional science is limited and wasteful. Radical ecology considers the method of Goethe. His natural philosophy incorporates a world view of organic dialectics; its methods are contemplative nonintervention and the primacy of the qualitative. Knowledge comes of itself, in leaps; a gestalt is perceived. Thought experiments and computer simulations can supplement observation. The qualitative cannot be subordinated to the quantitative. Qualities must be evaluated.

An alternative science considers every-day observations, unique occurrences, short-lived phenomena. Goethe recognized that different people are sensitive to different aspects of a thing. Any investigative effort should incorporate the observations of many others. .

To examine nature in general, we must shift to a taoistic approach, asking rather than telling, observing rather than manipulating; receptive and passive, not active and forceful; "nonintruding," and noncontrolling. It stresses observation rather than manipulation; it is receptive rather than forceful. Classical objectivity may be contrasted with taoist perception. Caring perception provides kinds of knowledge not available to remote researchers; this is especially true in ethological literature: Maslow cites his own work with monkeys; Lorenz, Tinbergen, Schaller, Van Lowick-Goodall, and Fox have found it to be true. This is the way a good therapist, teacher, scientist, parent, or friend functions.

What is necessary is not a primitive animism or a single-vision science, but a scientific animism, to understand our animalistic nature and use it as the foundation for a sound human ecology. A scientific animism would consider the relations of humans to vegetation and the human attitudes toward ecotypes, like open plains or dense forests; it would consider the need for sacred places, and open, quiet or wild landscapes; it would consider territoriality, aggression, and the aesthetic reaction to the wonder and beauty of life. Radical ecology is a scientific animism, a soul science (Anima is from the Latin word for soul). Nature is a feeling system. We need an animism to approach nature. This animism would allow us to behave "as if" nature were intelligent and sensitive, with the proper reverence.

A scientific animism would be concerned with far more than the anatomy and taxonomy of animals. It would be concerned with mutual experience between human and nonhuman animals, with the need for touch and the phylogenetic possibilities of animal empathy—dogs, for instance, exhibit strong physiological changes when they are petted and even human blood pressure drops.

Radical ecology considers the vast scope of ecology, including the economic and political behavior of human beings. The global character of its approach permits the creation of an ecological ethic and the realistic valuing of nature.

Recognize What Corporations are Really For

The first corporations were quite limited to public service, such as private highways, and did not challenge their host nation. As a result of changes in technology and several social changes, however, corporations were able to overcome limitations to their business. In 1886 the US Supreme Court made an incomplete ruling that corporations were persons, entitled to rights and privileges given to individuals by the Constitution and Bill of Rights. The corporation is legally fictitiously a person, an individual, although now they may be immensely large, rich, immoral, and powerful supercitizens.

Supreme Court Justice Louis Brandeis noted that US citizens were reluctant at first to grant corporations privileges for doing business, even though they were recognized as being more efficient. The reasons for this reluctance were: Fear of encroachment—on liberties and opportunities for individuals; fear of subjection of labor to capital; and, fear of monopoly. Privileges were granted, with restrictions on the size and scope of corporate activity. Gradually state governments removed those limitations. The corporate system evolved like a feudal system, according to Brandeis, where American society came to be ruled by a plutocracy of a few old, white men. Even more controls have been lost since the 1950s. The stockholders have lost control to management, thus ownership and control have become separated. Managers can pursue a course without the direct supervision of the owners. Antitrust laws failed to control corporations. Regulatory legislation as a legal limit on corporate power has also failed. The labor unions are no longer restraints on corporate power.

The public responsibilities of corporations, according to Harvard management, are to grow and prosper—thereby providing customer satisfaction, employment, taxes, and contributions to the economy—and to control their hazards. According to Milton Friedman, the only social responsibility of a corporation is to make money, by striving after profit as an efficient agent of production, although he admits that the corporation should conform to the rules and norms of society.

The US Constitution was written during a specific social context, at a certain level of technology, and at a certain point in the development of the industrial revolution. The forces that have developed as technology and corporate organization developed are not consistent with democracy and its ideals—and in fact, the constitutional and economic vision of that time have been rejected. The new system resembles an iceberg; the visible parts look like a democracy with a free market, but the invisible base determines where the ice heads, how it functions and who gets cool drinks. Because it was not foreseen by the US Constitution, this situation has escaped the traditional controls and limits. Corporate icebergs can direct the free market, by-pass democracy, and dominate people's lives. Corporations are powerful and indifferent to human life as well as to the free goods of nature, which is why not everyone gets lifted by the prosperity of the dark-side lords or why nature is also decaying. This new system has arisen from the alliance of government with private corporations.

Corporations have developed their own constitutions for operating for profits. Among managers there are shared knowledge and assumptions, acquired in schools and social situations. These assumptions become the real operating constitution of corporate government. Skills that will advance the interests of the corporate government are the ones that are selected, but understanding of the world below them or of the effects of their decisions on that world are not considered important. The shared assumptions include: Impersonal economic forces (a myth not often recognized) produce better choices than planning by thoughtful people; Economic growth is the measure of well-being of society as a whole and benefits everyone (another myth); All important social values are quantifiable and measurable (trust, loyalty, and beauty are outside but can be willed by strong people); And, the product of this system is the best system, not necessarily good people or a healthy society and environment. Based on these assumptions, the elite managers make decisions that affect all of society, although this consensus may not be conscious. The access to position and wealth has become controlled by corporations now.

Corporations are ruled from top down. Rules are not adopted democratically, or enforced with the fairness required by law. Employees obey the dictates of corporations to avoid being laid off. Employers then demand subservience and obedience far beyond what governments require with laws. And, the punishment, of loss of job, is very effective. Corporations have become exempt from the Bill of Rights, because the Bill applies only to actions by the government and its agencies. They do not allow employees freedom of speech or due process. Under the flag of efficiency, institutions have adopted an authoritarian model. A cowed workforce has advantages for a corporation. Civil liberties can be suspended in the name of profitability. The technology and communication works better in authoritarian mode. Its wealth gives it power. So much so that it does not require the use of force, although the public government allows it to use other forms of coercion.

Some efficiency is gained by corporate control of resources and employees. Corporations control the entire market. Thus, the market is not free; it is fixed by corporations. Prices are set by the costs of production, including bloated bureaucracies and executive salaries, rather than by demand and supply in the free marketplace. Price and production are controlled by planning, research, and programming, not by a free market. Economic power has been leveraged to political power, where corporations can deny free speech to employees outside of the work place. The economic government, according to Robert Reich, creates a corporate hierarchy, with pervasive inequality between all levels. To keep profit growing, corporations have been willing to accept damage and conflict.

In the past, corporations exploited child labor, until the government intervened. Corporations could dismiss employees without offering them assistance, so the government started to provide social security, and unemployment and training benefits. This response to the narrowing of opportunity and independence has been termed the 'welfare state.'

There are new kinds of tyranny from corporations. US citizens once fought taxation without representation, but they seem reluctant or indifferent to fight economic decisions without representation. Is corporate tyranny more acceptable than governmental? Is abuse of power by kings not acceptable, but abuse of power by corporations okay? Are unjust executions by government terrible, but cancer from toxic wastes from corporations acceptable? Is government mismanagement reprehensible, but abandonment of communities by capital just a part of business? Economic tyranny could also be reduced by balancing society more, by recognizing the immeasurable values that exist before being economized in an economic dimension. Control of corporations could lead to a better balance of things.

Although the government discouraged monopolies, it did nothing to regulate oligopolies—the control of the market by a few large companies. Corporations increase their power temporarily because of our failure to grasp their nature. The scale of corporations gives them more power. Management techniques, or technology, can augment power. We allow corporations to enjoy unprecedented benefits from a minimum of regulations, because we think that they are necessary to continue the growth we think we need to be prosperous and happy.

To be really successful, corporations could adapt a more comprehensive model, one that reflects stability, cooperation, justice, and respect for nature. Profit making is a necessary part of a for-profit business, but not the sole reason for business, or even the first, which is service. The best business serves public goods as well as private interests. A corporation provides a service, by employing people at decent wages and benefits—that is what it should be for.

Make Corporations Ecologically Responsible

Every corporation depends on the stability of the environment and on the stability of social institutions. The environment provides air, water, and land, and provides renewing (both physical and psychological). Institutions, from sanitation, police, schools, churches, and community centers, provide a supporting network. As these institutions wobble or fail, corporations may have to subsidize or replace them to survive. Corporations have at least three large, ecological responsibilities: First, to be economically healthy. The first responsibility of a corporation is the maintain its own health, to mature organically, limiting its size and impact to the locality. To do this, a corporation needs to *create a department* with ecological authority to envision long-range plans and impacts. *Plan all foreseeable consequences of a product. Adjust corporate strategies to changing values.* Smaller social and cultural groups have different and diverging values, so corporations are going to have to adjust to a diversity of values instead of to a monolithic standard. *Work to delineate a new information model* of production in which the stages of a process (capital, materials, workers, design, advertising, selling) are simultaneous and synthesized. *Enter partnerships with the employees. Promote the principle of least effort,* allowing the company to consume less, recycle, use longer, and avoid waste.

The second responsibility is to maintain the health of the natural communities—because environmental health is the basis for community health, and community health is the basis for economic health and worker health. *Be accountable for ecological impacts. Avoid interference with natural processes. Integrate loops and material flows;* internalize cycles. *Convert to ecological grounds practices. Promote ecological design,* which starts with questions. Is the product low- cost, aesthetically pleasing, and ecologically wise? Where does it fit in society? Ecological design, both responsible and socially responsible, must be radical, that is, rooted in a community in place. Membership in a place, in fact, leads to community. Corporations must become responsible members of the community.

The third is to support the health of human communities. It is hard to protect communities when the way most business is done tends to disrupt community life. Because of its size, power, and intentions (for profit), the corporation should take higher risks than the surrounding communities. *Support the community. Design the corporate structure* and size for the community. *Behave ethically.* An ecological corporation could use corporate buying power to promote acceptable technologies and discourage unacceptable practices. *Participate in the economic and social functioning of the community. Promote ecological education in a total context and interdependency.* And, *Implement community responsibility.* In education, integrate business with humanities; the responsibility for the welfare of the citizens belongs in the community, as does education, safety, and the whole infrastructure.

The corporation, regardless of its legal definition, is a long-lived, collective, impersonal body. Yet, it has more physical, legal and moral power than any one individual. Its investment is long-term in actuality. Many stockholders keep their investments for decades or a life-time. They are not concerned about only one dividend. Like the corporate organism, they want the long-term outlook to be positive; they want to know that their investment is stable and that the quality of life it encourages or supports is continuous.

The complexity of environmental problems should not permit escape of responsibility. The context of corporate responsibility falls within the spectrum from individual responsibility to social responsibility (the designation of property or trading conventions—capitalism or communism). Perhaps that responsibility could be enforced if the entire earth were incorporated and concerned with maximizing its own values: healthy beings in living contexts. Certainly not having `free' services and resources would force corporations to internalize all costs of production.

In any case, there are strategies that a corporation could pursue to become ecologically minded. Instead of treating decisions as trade-offs, an ecological corporation could aim at a congruence of moral, economic, and ecological objectives. Responsibility could be manifested in organizational structure, manufacturing, and marketing practices, without departing from economic decision making.

Such a corporation could bring corporate research and development capacity to bear on the transition to a balanced society. Where technologies play a role in the transition, companies can assume social responsibilities equal to their size and wealth. By commanding their vast resources, corporations can ease the transition to a better society, which would actually meet their needs for stability. The model of corporate life needs to change, from dependence on continuous growth, of profits and waste, to be being based on stability, cooperation, justice, and respect for nature.

This requires redesigning the economic model as well as the business model. Government can strengthen national economies, which are the basis for self-sufficiency and self-reliance, and control the global economies to make them more committed to social and ecological responsibilities. This may mean regulating global corporations so that they cannot take advantage of some resources and citizens. An ecological economic model shows how human welfare can be increased without growth or profit, using rules that limit drawdown or overshoot, or any of the other catastrophic trends that are deepening. It would slow resource depletion while increasing productivity and efficiency. It would reconnect processes into cycles of reuse and lower waste streams causing pollution and dead-end sinks. By understanding the limits of ecological and political systems, by respecting the properties of healthy ecosystems, place and cultures, ecological economics can still provide for the needs and luxuries of most all of humanity—allowing self-sufficient cultures to remain mostly outside of a new economic system.

Incorporate the Earth!

Political systems are impotent to stop the massive interference in ecosystems by international corporations. The simplest and most direct way to give the earth a voice in the development of the earth by humanity is to incorporate the earth following international law. The entire planet, with its biochemical cycles and nonhuman communities, would become one legal body. Since corporations are human constructs, however, humans would have to represent ecosystems and their wealth of living organisms.

Corporations are recent devices created by states for public purposes. Most early American corporations, for example, were concerned with travel (turnpikes and inland waterways) or safety (fire insurance)—they resembled public agencies more than profit-seeking associations. In fact, the exclusive privileges and political power granted to corporations were based on the implicit promise of social services.

Changes in societies, from rural to urban, from sparsely to densely populated, from culturally diverse to monotone, have transformed corporations and the societies themselves. Business corporations now provide the bulk of goods and services in many states. The scale of these corporations, the processes of production, and the size and needs of human populations, have altered and degraded many ecosystems and biogeochemical cycles.

Successful modern corporations create an identity based on their purpose in providing goods or services; they define their business in terms of profitability, growth rate, cash flow, and competitive position; they develop a corporate vision, with specific objectives and strategies, including long-term vision, collection of ideas and creative implementation, aggressive manufacturing, and reliable finance.

The purpose of a corporation often transcends simple financial gain—the corporation seeks to maintain its own existence, before profit. Financial objectives (sales, assets, profits) exist to sustain its existence. The goals that most motivate corporate managers are survival, independence, self-sufficiency, and self-fulfillment. Yet, these motives are consistent with the financial objectives of the corporation: to maximize corporate wealth. The responsibility of managers is to maximize the value of the company. Furthermore, because corporations are long-lived, that value should last a long time—a good reason for looking beyond the ten-year monetary horizon and the lives of its managers.

Although current wisdom, from Milton Friedman et al., holds that a corporation's only responsibility is to its stockholders, corporations are being pushed to include social purpose in their strategies, again. Alas, they are doing poorly at it. They do not know how much responsibility to take, or where to put limits, or whether to pursue policies that diminish their profits. Corporations have proved spotty in doing social and environmental good. It would be more appropriate to have them deal with the environment as a corporate entity concerned with maximizing its own values. Of course, that would mean no more "free" resources or environmental services.

The important advantages to incorporating the earth are the same as for incorporating a business.
1. Managerial flexibility: the stockholders are separate from managers; responsibilities are assigned by needs of the corporation.
2. Limited liability: the corporation borrows and repays. It shields its members from hazards to which they would otherwise be exposed.
3. Financial advantage: the ownership of assets can benefit stockholders and the corporation.
4. Tax advantage: investments in the good of the corporation may not be taxed by nations.
5. Estate planning and longevity: the corporation exists indefinitely beyond the lives of its participants.
6. Central management and representation: a large and complex business needs operational and managerial efficiency. Many of the participants have no direct voice in the operation—they must be represented.

The earth incorporated would focus on a core business: to ensure the integrity and continuity of life and all its connections and to secure the opportunity for development free from undue interference. It would operate to optimize values, like any good corporation, but the values would be ecosystem values (fungus values and earthworm values, as well as human values).

A temporary Board of Directors (the undersigned) would adopt bylaws, elect working officers, approve stock certificates, open accounts, and arrange a stockholders meeting. The stockholders would elect new directors, possibly from United Nations representatives or directly from elections, and decide on dividend declarations.

Stockholders, as citizens of independent nations, would turn over common and national property to the Earth Corporation, which would issue stock certificates to the stockholders. The corporation would allocate the purchase price of stock to capital at par value. Most of the shares—the percentage to be determined by the board as necessary to the operation of ecosystems—would be treasury shares. Anything more than par value would go to capital surplus, and only capital surplus could be distributed as dividends. Stockholders have the right to receive these dividends equitably, without resort to traditional distributions of wealth.

Stock certificates denote ownership of the corporation. Although the stockholders own the corporation, they do not own the property of the corporation, the earth, which is owned by the corporation itself. Stockholders, as individuals, groups, or nations, could make agreements about how business would be conducted, about what resources would be used or traded.

The elected board of directors would make decisions of distribution and limitation. Percentages would be deducted from the interest for the operation of the corporation and for equitable distribution to nations less favored by chance with biological or geological wealth. Furthermore, since the dividends would be distributed among people according to net ecosystem productivity and resource availability, no advantage would be gained by nations having large populations.

The basic functioning system would be considered capital, thus limiting the human use of resources and probably the size of human populations. Interest would accrue in the form of net ecosystem productivity and diverted percentages of materials, such as gold or water.

The earth incorporated would solve the problem of having to value ecosystems in monetary or quantifiable terms; its systems would be untouchable capital. The human value of resources like copper, air, or water would be equated to the technological cost of recycling or producing them.

Raw material and energy are only two facets of the capital of a corporation—another is human ingenuity. Thus, human wealth would not be limited by restrictions on the availability of resources, but rather by a shortage of ingenuity.

An incorporated earth would be instrumental in conditioning international corporations to their social responsibility and in internalizing all costs. This corporation and governments could use traditional means, such as credit access, low interest rates, and setting priorities on equity issues, to evoke public interest in smaller and healthier human endeavors.

Articles of incorporation would name the corporation 'The Earth Inc.' and delineate the purposes, such as protection, maintenance, and conduct. Limits would also be set out. The address and board of directors would be elected.

> *Author's note: Four of us signed this paper in 1984 and tried to get it incorporated by the State of Washington. We were told it could only be incorporated as a name of a state corporation that would have no power over any territory. We declined to continue. Later, we were unable to get UN attention even at the local UN group. Likely, the planet can only be incorporated at an international level with the agreement of all nations.*

Perhaps the form of a nonprofit corporation is not the proper approach to protect regions. Perhaps, a region could be represented well by some sort of legal trust, as are private properties. This might solve the dilemma of ambi-human species as well as future human generations, which require a much longer time frame than most plans.

Bali's water-sharing temple system for rice farmers is a good example of commons management. And, there are other examples of this kind of management of limited resources. The Spanish Huerta was also a system for distributing water and resolving disputes.

Garret Hardin, after criticism of his article on the "Tragedy of the Commons," pointed out that the problem was not common ownership as much so open access to a common without the limits of social structures or rules. Ocean fisheries and the planetary atmosphere are two examples of modern commons with limited restrictions. Tragedy occurs especially when social structure breaks down or when the scale increases beyond the control of any local organizations, as with atmospheres and oceans. Overpopulation can put pressure on the commons. External corporations can put even more pressure on a regional or global commons.

The commons can be productive where there is a common culture with rules and laws. Switzerland and its alpine pastures, or the rice fields of the Philippines, are examples. The commons of the ocean and atmosphere are too large for local control, however. These real global commons have to be ruled by a global institution. This is possible because the management can be scaled up to an international body. Individual people and cultures have to recognize that the commons provides for all equally and have to have rules.

Possibly, commons could be managed in a Trust. In common law legal systems, a trust is an arrangement whereby property, including real, tangible and intangible property, is managed by one person, or persons or organizations, for the benefit of another. A trust is created by a settler (or trustor, grantor, donor, or creator), who entrusts some or all of his or her property to people of his choice, the trustees. The trustees hold legal title to the trust property (or trust corpus), but they are obliged to hold the property for the benefit of one or more individuals or organizations as the beneficiary (a.k.a. *cestui que use or cestui que* trust), usually specified by the settler. The trustees owe a fiduciary duty to the beneficiaries, who are the "beneficial" owners of the trust property.

The definition of a trust allows it to maintain an asset for future beneficiaries. For example, the Pacific Forest Trust protects private forests from clearcutting and development, through conservation easements that limit the kind of use that might harm the ecosystem. Private landowners can harvest some trees sustainably. Marin Agricultural Land Trust buys development rights to farmlands. Oregon Water Trust restores water flow to endangered streams by acquiring water rights. Could it apply to oceans? In each of these cases the owners hold the land and can benefit, within limits, from it. Would this work with nonhuman owners like the planet? Can local commons management be applied to global commons to address global problems, such as ocean and atmosphere?

The trust is governed by the terms of the trust document, which is usually written and in deed form. It is also governed by local law, although it could be governed by some new global law. There are a few basic principles for a trust: Property of any sort can be held on trust; the uses of trusts are many and varied.

Trusts can be created by written document (express trusts) or they can be created by implication (implied trusts). On a regional scale, the trust would be created by one of the following: A written trust document created by the settler and signed by both the settler and the trustees, or a court order. Due to the legal limitations of the ability of nonhuman species to communicate legally, a court order would be the best way to set up a regional Trust for Bali.

There are formalities for establishing a trust. The property subject to the trust must be clearly identified. The beneficiaries of the trust must be clearly identified, or at least be ascertainable. In the case of discretionary trusts, where the trustees have power to decide who the beneficiaries will be, the settler must have described a clear class of beneficiaries. For the regional trust, beneficiaries could include any living beings alive or not born at the date of the trust. Alternatively, the object of a trust could be a charitable purpose to be held by the United Nations or organization rather than specific beneficiaries.

A Trustee can be either a person or a legal entity such as a company. If there is no trustee, whoever has title to the trust property will be considered the trustee—in this case a global governing body. Otherwise, a court may appoint a trustee. Trustees are nearly always appointed in the document (instrument) that creates the trust. The Trustee, especially for a regional Trust, has a huge responsibility. She may be held personally liable for any issues that arise with the trust. However, due to the differences between the values of the Trustee and the planet, the Trustee has to be overseen by a regional or global governing committee.

The trustees are the legal owners of the trust's property. The trustees administer all of the affairs attendant to the trust. This includes investing the assets of the trust, insuring trust property is preserved and productive for the beneficiaries, accounting for and reporting periodically to the beneficiaries concerning all transactions associated with trust property, filing any required legal documents on behalf of the trust, and many other administrative duties.

By default, being a trustee is an unpaid job. However, in modern times trustees are often lawyers or other professionals who unwilling to work for free. Therefore, often a trust document would state specifically that trustees are entitled to reasonable payment for their work—perhaps an amount equal to the average income of a mid-range position. The beneficiaries would be the equitable owners of the trust property. Either immediately or eventually, they would receive income or benefit from the trust.

The regional trust for Bali could have elements of a constructive trust, since it would be imposed by law as an "equitable remedy" against those holding the assets as a matter of luck or discrimination. It would have elements of a spendthrift trust, since humanity is unable to control its spending of natural capital; the trustee, perhaps an agency of the United Nations, would have the power to spend only ecological interest. It would resemble a unit trust, in the sense that all human beings would possess a certain share of the interest of the planet—of course, nations would control a percentage, determined by the ecological carrying capacity, which would be divided equally among the population; this would encourage nations to normalize their populations. It might resemble a public trust, in the sense that it would have the object of keeping the planet healthy, as the source of most capital.

Co-ownership, as a trust, could be divided between species first—that is, the ownership of the home, earth, is shared by all living beings, and all living beings should have some legal representation. For Bali, living water would have an interest. If it were a hybrid trust, the amounts of the trust interest could be paid out at the discretion of the trustee; this might be used to settle long-standing grievances and inequities. It could resemble an incentive trust, in part, because it would encourage some behaviors, such as inventiveness and frugality, and discourage others, such as waste or inequity.

Obviously, there would be many benefits to setting Bali up as a trust. The greatest benefit would be legal protection of many areas for ambihuman species. In fact that would be a major purpose of a trust. A second major purpose would be to equalize the income from global interest so that it would be divided equally among national residents.

Figure. Terraces in Bali
(Credit: Amy Ihrke).

We are Interfering with Nature

Nature changes as its parts interact. There are parallels between the interactions of processes, of animals and of humans in ecological systems. Science has concentrated on disturbance and exploitation behavior, but these can be contrasted with the interference behavior that characterizes the nonecological activities of the dominant industrial culture. Examples of each can be noted in wild ecosystems, forestry, animal cultures, archaic human cultures, and industrial culture. The word interactions is used, instead of words like 'events' or 'catastrophes,' to describe the feedback and cyclic nature of actions.

Humanity is exploiting nature recklessly, without attention to the minimal health of ecosystems. Many ecologists, such as Eugene Odum, have observed that complex communities have existed for thousands of years in relatively stable environments, even though these environments are characterized by regular disturbance and constant exploitation. These environments are now vulnerable to human interference, which is a different thing from disturbance or exploitation.

Disturbance, by definition, is an event that can be caused by climate, biological entities, or other actors. Disturbance is what causes change in an ecosystem. On a small scale, a single tree falling over is a disturbance. Although an individual dies, species continue. Mortality is a normal part of the life cycle of the forest. The disturbance may be necessary for the ecosystem to continue to mature; for example, according to David Perry, without windthrown spruce that expose mineral soil seedbeds, the northern forest ecosystem would shift to bogs.

Disturbances, if regular enough, become a regular feature of the ecosystem. In Florida, some species, such as Cypress, need the complete inundation provided by hurricanes to remain healthy. Yet, even catastrophic disturbances like hurricanes rarely damage more than 5 percent of a forest;

Exploitation is the normal use of a resource or of a species by another species, including the human species (this ecological definition differs from a sociological definition, which means 'selfish or unethical use,' although the ecological use may suffer from negative connotations from the sociological use); in fact, ecological exploitation has a rejuvenating effect on populations. Exploitation is contrasted with interference, an activity that can degrade, destabilize, or destroy entire ecosystems.

Interference is not a form of disturbance, exploitation, or competition; it is destruction without gain to any species; sometimes it is caused by planetary events, but in the case of human interference, it is the destruction of the structures and processes of evolution for large-scale, one-species, short-term economic gain. The pandominance of ecosystems by humanity is related to the biological and cultural characteristics of the species; this should be recognized as a cause of interference. Ignorance and indifference are identified as major reasons for continued interference.

Interference has been a rare phenomenon on earthly ecosystems; it has happened in the past as the result of global catastrophes, such as meteor impacts. Now, interference, as opposed to more limited and predictable disturbances or exploitations, is threatening the stability of all ecosystems. It is dangerous to interfere with the processes of ecosystems because it disrupts the communities on which other species, and ultimately human communities, depend. Furthermore, in the deepest sense, it violates the idea of living together with other species on the planet. The proper relationship of humanity with nature includes competition and exploitation and mutualism, but not interference.

Practice Noninterference

The basis of all value is being (the verb form of the word). Every being has value in itself by virtue of its existence. This is reality undistorted by human needs. Of course, humanity can still enhance nature with its presence, in a less reckless or interfering manner.

The attitude towards beings as they are can be expressed as an ecological rule of noninterference, which is a basis for appropriate ways of balanced exploitation, so that human and other communities can all thrive. The rule of noninterference means 'letting be' in the words of Martin Heidegger), 'letting alone' according to E.O. Wilson, and 'not killing for pleasure' as Michael W. Fox commands. Noninterference is not indifference, which is diffuse. It is caring. Noninterference will not lead to chaos, poverty, and stagnation. The technocratic vision strives for 'life under control,' but the earth is self-managing, productive, efficient, and orderly. Life does not need to be always under control.

Noninterference can be derived from nonviolence, or from taoistic nondoing. To examine organisms, and nature in general, we must shift to a waiting approach, asking rather than telling, observing rather than manipulating; being receptive and passive, more than active and forceful; nonintruding instead of invasive, noncontrolling instead of manipulative. This concept of ecology as patient practice stresses noninterfering observation rather than controlling manipulation, reception rather than force.

We need to practice the rule of noninterference so that all beings can exist in place. This attitude would entail using what is necessary, exploiting parts of some ecosystems, perhaps all of a few others and none of some, changing some places to fit human aspirations, and killing some plants and animals for sustenance. But it would also mean limiting humanity and its technological effects, limiting human use to local impacts, and letting other beings live without interference. It is not necessary to dominate or terraform the ecosystem completely to save it.

A noninterference approach to ecosystem management, the essence of a taoist way, is to let the system take its own course. Therefore, once temporary constructs were in place, whether planting or cutting or any other manipulation, the system would be allowed to develop with minimal intercession.

Formal management, based on the rule of noninterference, weaves people back into the fabric that supports them and in a sense makes them subject to the constraints of ecosystem processes. With so many agricultural, domesticated, and wild systems needed to feed, cloth and house people, formal rules may be necessary. First, we must allow natural processes to operate in mostly natural systems. Next, we have to stop simplifying the systems and restore the critical species where known. In agricultural systems, we need to manage the system with minimum subsidies, especially in terms of pesticides, fertilizers, and energy use—there are working alternatives to these industrial methods. We can align our activities to natural processes to take advantage of them. We also need to manage activities that could upset equilibria in near-wild (e.g., forest plantations) and wild systems. We also need to restore the wild context of artificial systems. According to Garrett Hardin, many of these ideas necessary for fitting humanity into the patterns of nature are known, but not very popular, or against the vested interests of industry, so that they do not get implemented. Eventually, however—that is, now—we will have to fit our civilization into the limits of natural communities and of the planet as a whole. And, the rule of noninterference can help make that possible.

Figure. All theft?

Stop Stealing!

Even low average levels of food and fulfillment can be maintained only through theft from other species and theft from future human generations, and through the degradation of billions of humans as well as the ecosystems on which they depend.

How many die directly from hunger, since hunger leads to disease, such as pneumonia, and disease gets the credit; how many die from unclean water, since the contaminant or the accident gets the credit? It is also difficult to calculate deaths from theft, economic dislocation, lack of planning, bad designs, or depression. Cause and effect make a complex dance of candidates in a network of things. Only specific diseases, accidents or bullets seem to be unambiguously fatal, and even these are often causally linked with many other contributing factors.

Ghost acreage is the additional land, from sources outside a nation through trade, theft or conquest, that a nation needs to supply the total amount of food and fuel.

We have allowed the thefts of life, intelligence identity, and choice to negate any gain from cleverness or bigness. In a mass consumption society, people impoverish themselves spiritually while impoverishing others materially. This is theft. As with the Christian ten commandments, most loss can be reduced to theft, whether of a life, mate or name. Most of our modern problems can be considered the consequences of forms of theft, such as of life, common sense, and choice.

The lowland forests of Malaysia, Indonesia and the Philippines are being ravaged the fastest. The rape of the tropics is endangering hundreds of thousands of species. Over 25,000 known plant species and over 1,000 species of mammals, birds, reptiles, amphibians, and fish are threatened with extinction, as their habitats are eliminated. Overfishing is destroying the fisheries' support systems. In the US, losses to fisheries from shore "improvement" and degradation cost $86 million a year.

The relationship of humans to humans has also changed in the past thousands of years. Human lives are stolen, not only through war, but large-scale murder, as well as through diminution of human value and denial of resources.

Theft of common sense. Consumers retain little more than a dim notion of the past. Universities report a lack of interest in events that occurred before the current year's athletic season. The sense of time falls in upon itself, collapsing like an accordion into the present. Knowing nothing of history and expecting nothing of the future, people cannot escape the fearful isolation of the present. They join together in a melancholy herd, clutching at everything, but holding nothing fast. Our growing emotional and intellectual detachment is the greatest threat to nature. Heroic narcissism has replaced nature with humanity; nature no longer provides the mirror to reflect human aspirations, a television screen does.

Theft of choice. Unbridled consumption in a laissez-fare economy for fifty more years will probably carry humanity beyond a point of no return, leaving industrial society with insufficient resources to maintain itself and insufficient flexibility to retract. If lack of planning permits rapid increase in population and consumption, then our options may be diminished to two: A nasty, overplanned existence, or a squalid collapse.

The stresses from losses have resulted thefts, as attempts to balance or correct the situations. The very size and impact of humanity has made theft the only easy option, much easier in the short run than planning or self-restraint. Stopping a theft can be as simple as stopping a thief. But, theft has become such a complicated thing, many steps removed from the people who make the decisions and from those who carry them out.

Living with the Success of Others

At a time when every bloviated Luckster (a person who was lucky in timing, invention, theft, or presentation of an idea, program or plan of a consumer object) claims complete credit for their 'thing' and their subsequent wealth, the *same error of thought* is being applied to people who have 'failed' temporarily or permanently—surely it is their own fault, the Luckster judges, and all they have to do is pull themselves up by their bookstraps (something surprisingly rare as part of our apparel). Furthermore, we ought to deny them any help that might take money from anyone else's pockets (or taxes).

But, what if a new idea is ahead of its time? What if the invention has no context yet in the industrial world? What if the idea demands too much effort or thought from the public? What if they were victims of theft? What if they just had bad luck?

We seem to forget that most things are developed on the foundation of an advanced civilization, with electricity and roads, a postal service and public libraries. We tend to ignore the personal support group of many inventors and programmers, not just from spouses and family but from institutions, from businesses, corporations, universities, and foundations. Most everything that is thought of and copyrighted or whatnot has depended on a tremendous support network—and that does not even include pets, indoor plumbing and space heaters.

We should not ever be giving exclusive credit to anyone profiting from all this support. Neither should we be blaming those who lost their jobs and were thrown into a cheap labor market with overpriced and refined skills. Or those who have been beaten down by bad luck, conspiracy, divorce, or circumstance.

Those people need some time to find appropriate jobs; they deserve help because they paid into the support system while they were employed or working. They do not need to be humiliated by the institutions set up to help them, or teased by politicians who can vote their own pay increases or permanent insurance. They do not need to be made fun of by talking heads in the newsreaders hours on television.

Let's please wake up and stop all this counterproductive blaming and screaming. Very few people have done anything on their own and very few people have chosen to be unemployed, homeless or addicted. And, very few people have gotten rich on their own; many have ridden the tide of dishonest profiteering from illegal and unethical financial practices, especially related to real estate bundles or microtemporal money transfers, and others have benefited from various bubble-like ponzi schemes. Let's understand where credit or blame lies, and not praise or condemn every individual who rose or sank with the tide.

Let Local Wisdom Temper Speed & Global Power

There may be a dramatic redistribution of power, from slow countries to fast, Alvin Toffler foretells (in a perfect echo of the industrial-financial Zeitgeist). He states that, historically, power has shifted "from the slow to the fast," whether speaking of "species or nations." Certainly, being faster to the industrial market has advantages for many international corporations. But, this kind of speed is not applicable to species. Slow species have survived as well as fast, either adaptively or neurally; many fast dinosaurs perished before their slower mammalian contemporaries, for instance. For species, size and flexibility seem to be more critical for survival than speed.

Then, Toffler notes that the industrial revolution stepped up the metabolism of economies, but does not seem to make any distinction between good or bad metabolism (fever as well as excitement speeds up a metabolism). Truly, we are speeding up our use of resources without knowing where they are coming from or going to. Modern economies, embracing the idea that "nature is capital," draw on the accumulated "capital" of ecosystems for production. By ignoring the real cost of the capital, as well as the costs of natural services, such as nutrient recycling, soil building, and atmospheric renewal, these economics create a temporary wealth (similar to the healthy flush of a fever, perhaps) and a long-term imbalance. When an economy falls out of balance with its local environment, massive disruption often results; industrial economies have only avoided disruption by trading advantageously with other economies, by using fossil fuels, and by promoting institutional inequality.

Continuing his paean to speed, Toffler states that fast economies generate wealth and power faster than slow ones. But, what kind of wealth? Financial or cultural, agricultural or symbolic? And, what kind of power? Mechanical or organic, political or personal? Industrial economic wealth is merely a small part of the wealth of the earth and humanity, most of which has little value to that economy.

There seems to be an acceleration effect that makes each unit of time saved more valuable than the last, creating a positive feedback loop—inadvertently identifying the archetypal problem of modern economics—runaway positive feedback loops leading to catastrophe. The fast economy he describes seems to depend on fleets of hypersonic jets racing around the world with the elite and their tonnage of possessions. Telecommunications, transportation, and tourism all accelerate, blithely unaware of their impacts on family structures, biogeochemical cycles, including the ozone layer, and wilderness. We have not learned anything about the waste of speed.

Toffler foresees the emergence of an electronic neural system for a global economy, without which any nation will be doomed to backwardness. What kind of backwardness? Lack of fast things? Lack of professional enslavement? Lack of art, play, or culture? Lack of food, tradition, freedom, or happiness? He describes the fast economies that are forming and concludes that slow economies will have to speed up their responses or risk becoming uncoupled from the fast lane. It might be good for countries to be uncoupled. Uncoupling economically might be a sound option for traditional societies unwilling to make the same mistakes as industrial ones. Local communities are based on traditional cultures, which have long-term lasting power. Traditional cultures often have wealth-leveling properties, absolute property ceilings, fixed wants, and production coupled with need—all of which results in a stable economy. Then, efficiency and productivity are less important than use and appropriateness.

The current 'mood' is that the nonindustrial countries are faced with a shortage of economically-relevant knowledge. Are they? What kind? The knowledge of how to find or grow edible and medicinal plants? The knowledge of how to make appropriate houses and cooking utensils? Toffler touts knowledge-based agriculture as a cutting edge of economic advance; how knowledgeable can it be, if it ignores erosion and beneficial insects?

Traditional communities have lost more knowledge than we will have in the near future. What happened to our rich biological knowledge of animals and plants, to our rich mythical knowledge? Economic success not primary; it is secondary, as is money, the accumulation of goods, and prestige. We are accomplished in the primary meanings of life. The satisfactions from being in a culture in place, from planting trees, growing apples, watching birds, playing with children, and making love are primary. They are not speed-dependent. We lack the wisdom to act as if we believed this. Is fast technology a necessary part of happiness? Those who are uncomfortable with primary meanings tend to become addicted to speed, power and possession, as a frantic way to avoid awareness, silence, or responsibility, as a replacement for being grounded.

Nature provides the source, of wonder, of the sacred, of otherness, and of the wild. Humanizing the world has made it tedious, uniform, and dull. *Economics is dull! Toffler's assumptions are dull.* The needs he describes are transitive wants, and their only measurement is quantitative. For fertile nature, we have substituted a sterile model of production and economy. The model is reductive: Trees become resources, people become labor. More is more, faster is better. Although speed is our normal response to dullness, the celebration of speed for itself is ultimately unsatisfying.

What is the result of our fascination with speed in everything? Dismissing nature in disgust, we attempt transcendence through speed. We speed away from nature, from our own bodies, and base our civilization on that momentum, praying, requiring, that it never stops. People's souls die, but secure in their power, they manage the things of civilization and inhabit the treeless flatscapes of the malls of commerce, comforted by the banishment of wilderness and the capture of animals in zoos and of free people in reservations, satisfied that their young are mercilessly tied to televisions and computers, acquiring information without touch and speed without grace.

Nations and communities do not all have to follow the same path and the same rules at the same time and at the same rate. Cultural success is not the "survival of the fastest" any more than it is of the biggest or shallowest or newest. Perhaps if we remain unconscious, there will be a power shift to the fastest that will homogenize and level human cultures. But, we can consciously imagine alternatives and work to preserve cultural and natural diversity and the richness of existence.

We have the knowledge to save cultures, to restore places, to participate in the cycles of the earth, but extra speed and power are not required. The pace of nature is generally balanced and well-established; we violate it at our risk. If we adjust to the pace of the growth of trees and to the movements of animals, we would not be risking extinctions and famines, shortages of water and fuel wood, and the death of humaneness.

We do not need to give our power to faster economies. We need to shift power to local communities through self-reliance and participation. A community protects individual freedoms, guards regional culture, and holds groups accountable for their use of power. In communities, people can decide to be conservative or to grow and gamble on innovation. Communities can have different economic attitudes, paces, and goals. A community that is balanced and flexible, in tune with natural cycles, based on traditional values—in which industrial production is limited to appropriate goods—can absorb the shocks of change far better than a powerful, accelerating, postindustrial, national vehicle.

Another Proposal to Empower the United Nations

The notion of a world government seems to satisfy a basic human craving for unity and order. And, an implicit world system is evolving through economics and science. A global order is necessary to govern the system, but, at the current stage of international relations, there seems to be no agreeable path toward a benevolent world order. The partial adoption of international institutions is insufficient for a world order, especially if those bodies are only advisory.

The United Nations (UN) is the only existing body with the machinery for constructing a world order; the beginnings of a comprehensive politics can be found in the special services of the UN: UNESCO, FAO, WHO, and the various technical aid services. As long as ecological and political problems are addressed in a framework of nationalism and military power, however, these organizations are treated as peripheral and impotent.

As it is structured, however, the UN is not capable of handling the responsibility for world order. For example, the UN's solution to economic problems is 'sustainable development' within environmental constraints. The Bruntland Report, which proposed that solution, indicated a five to ten-fold increase in world industrial output within the next one hundred years before population stabilization. While the appeal to growth is unarguable, it is really not likely to be sustainable in any meaning of the word, since this kind of growth does not recognize or respect known ecological limits. Considering that the current level of industrial output has imposed severe threats on human society and environmental health, even a five-fold increase should be able to destroy the cultural and ecological diversity of the planet.

Other actions of the UN, such as restricting membership in the security council to 'great' powers with nuclear arsenals, or its use of the veto principle, indicate that the organization has been captured by the status quo. Furthermore, even when the UN does make good recommendations, it does not have the power to coerce any nation to follow them.

Rather than replace the UN with a new construct, we must revise it. The UN has been a half-hearted investment, but it has historical appeal and wide support. It is a nascent global order, but it must be reorganized and empowered, as first advocated by the Jackson Report (1969) and the Hammarskjold Report (1975). These recommendations are just a continuation and emphasis.

The United Nations, an elected body, shall have the regulatory powers necessary to maintain a healthy global environment. It shall have advisory and regulatory powers to maintain the independence and integrity of its constituent nations and their peoples and places. It shall have the regulatory and punitive powers to rectify resource and human rights infringements. Only this body shall have international police powers and large-scale weapons. Various advisory bodies shall recommend policies and actions to nations. The UN shall have several basic functions: To Ensure a diverse and healthy biosphere; To Ensure a Diverse and Healthy Biosphere by creating categories of landscapes, from wild foundation to artificial ones; To manage common resources by forming new institutions, to set standards and redress inequities as well as maintain reserves of minerals and food; to protect unique human cultures: to coordinate the representation of cultures; to provide services to nations and individuals, especially education and security.

The UN would call for immediate action to combat catastrophes. It would implement emergency measures. Social changes can occur very rapidly, however, when the time is right for them; for instance, oil-producing nations became the financial equals of industrial countries within months. An immediate, realistic, coordinated program of action is needed, capable of being implemented by communities and global agencies.

The application must be *immediate*. The crisis of exponential growth and destruction cannot be solved just after some final limit is approached or passed. The crisis of ignorance cannot be solved by hurrying ahead and creating more problems. Immediate social reforms, the reallocation of resources, and the preservation of wilderness are necessary, because of the nature of the problem; we cannot predict global climatic or ecosystemic catastrophes. Substantive change and research cannot be delayed until academic controversies are resolved.

The transformation must be complete; it cannot be done partially. Global political and economic institutions must all be changed. The UN must have authority for the preservation of nature and human cultures. Holistic change will permit the reorientation and balance of local institutions.

Five steps are necessary: The transfer of military powers; complete disarmament of nations, except for police forces; catastrophic measures to stabilize and protect wild and human ecosystems; and, redesigning economies and polities. That which has been hitherto left unsaid-what we want to become, what we could become-could become explicit. Now is the time to define goals in terms of population, quality of life, and preservation of biomes. Goals are not some final state reached once for all time—they are a horizon. The UN offers continuity towards the goals.

We need meaningful ideas. Our attitudes and feelings toward nature need to be revitalized with evocative metaphors that let us accept responsibility for the part of the earth that we build, namely human culture and human landscapes. The truths of our unique cultures and the wild earth are apprehended through myths. The poetic language of mythology can fit all the facts and values, things and images, into our hearts so that we can feel them and act upon them—so that we can make good places. An empowered UN can provide the order to accomplish this.

Save Common Areas: The Palouse Ecoregion

Most wildernesses are limited to coniferous forests in remote montane areas, yet unspoiled common places, like grasslands, are being altered rapidly by industrial agriculture and human population pressures. Grasslands are among the most poorly protected biogeographical provinces.

The Palouse grassland is a geographic region of approximately 6 million hectares centered in Southeastern Washington. Its origin, topography, and soil composition are unique. Wind-deposited loesses form steep, rolling, dune-like hills that overlie the Columbia River basalts. The primeval vegetation was composed of dense stands of perennial bunch-grasses.

Dry-land farming has almost completely replaced the original vegetation, although fragments can be found in fence corners, right-of-ways, cemeteries, and inaccessible slopes. In spite of its uniqueness, there has been no successful attempt to save more than patches of the original vegetation. In the 1960s the Idaho Association of Soil and Water Conservation called for the expansion of the Great Plains Conservation Program to include the Northwest prairie. This, and later resolutions, were defeated for regional or financial reasons. Small research natural areas (RNAs of 10-15 ha) have been saved by Washington State University (WSU) as research areas, but no large stands of native grasses remain.

Many of the plant associations in the Columbia Basin Province are not represented in any of the small RNAs in Washington or Idaho, although some of them would be included in proposed areas. In the Zonal Meadow Steppe Association, the Idaho fescue/Nootka rose community has no representation (it occurs in Hells Canyon) and the Idaho fescue/Snowberry community is only partially represented (in the Kramer Biological Study Area). Most of the communities, such as Bluebunch wheatgrass/Idaho fescue, in the Zonal Meadow Steppe Association are only partially represented. Poor representation is one reason why large reserves are needed. Furthermore, the RNAs are not large enough to save viable mammalian populations, and this is another reason why large reserves are needed.

This reserve proposal recognizes that the desired size of the preserve is a complex function of the area's key species, quantity of suitable habitat, and minimum viable numbers of species. Large-bodied vertebrate species tend to have lower population densities, thus a reserve with large-bodied vertebrate populations will likely be adequate for herbivores, insectivores, and primary producers. The key mammal species in the Palouse are coyote, badger, and mice, with white-tail Deer as regular visitors. Determining the minimum number of individuals in a population to guarantee a high probability of survival results in widely varying minimum areas, depending on the key species selected. Using coyotes as the key species, the minimum area for the reserve becomes 1.05 million hectares. With a home range of 250 ha (2.5 km^2), the minimum area for deer would be 200,000 hectares. Usually, large carnivores are a sensitive indicator of carrying capacity.

The shape and size of a Palouse Reserve is determined by habitat studies of the unique natural history and conditions. The key plant species to be protected are Idaho fescue and Snowberry, with all their ecological relationships to micro-organisms and arthropods. The size should be large enough so that species will not be vulnerable to "extinction vortices," caused by genetic or environmental stochasticity. In this reserve, disturbance from farming, grazing, or recreation would probably be the greatest threat.

The recommendation for a Palouse reserve is 1 large area, about 200,000 hectares, buffered doubly by rehabilitated fields and then by a greater amount of fallow agricultural land, 3 areas of 3,000-10,000 hectares, and 22 satellite areas of 8-25 hectares, which would probably not be buffered. The reserve is divided into over 25 unequal areas. The largest portion of the reserve proposal is sited on the eastern edge of the Bluebunch wheatgrass province and includes at least five habitat types. Saving a million hectares would be economically or politically difficult under the current industrial monolith; 1 million hectares is about 17% of a region that profits immensely from growing grains and legumes. The 200,000 hectare figure is over 3%. As a percentage, this should be possible; it is less than a sales tax or an income tax percentage. In a truly eutopian society, characterized by rational planning and realigned priorities, 3 million acres would be set aside. But, such a vision can only be created with radical changes in political and economic regimes.

The large areas of the proposed reserve would be laid out on the SW/NE axis with large fingers extending to the NW and SE, to maximize the number of protected NE slopes. Since the winds are predominately from the SW, herbicide drift would be minimized. A range of elevations across areas would minimize the effects of climactic change—and the possibility of extreme change is rarely considered in wilderness design. Soils, drainage, and land-use history and ownership would also receive similar considerations. This would allow management for diversity on three different scales: Community, habitat, and region. The reserve would extend into the ecotones separating grassland and forest provinces; the edge effect would benefit many species.

In the Palouse, there are probably no natural landscapes and very few near-natural; most are semi-natural (pastures as a consequence of human activity) or artificial (totally humanized). The Palouse could not be restored to its original state, since many species are extinct, but it could be rehabilitated. There are a number of methods (such as direct planting into de-sodded ground) that can be used to restore species-rich grasslands. The inner buffer zone, being rehabilitated, would be the most expensive to create. The outermost buffer would be managed by benign neglect, despite the fact that others do not list this as a management option. Natural processes, such as fire, wind, or species explosions, would be allowed to operate freely, even if they altered the functioning of the communities. In fact, a reserve, in contrast to a preserve, is simply an area placed off-limits to human activity, where natural processes would be allowed to proceed naturally.

The cost of reserves would be high. Palouse land sells for $2,000 per hectare. The cost of a reserve, for restoration, has been estimated at $20-140,000 per hectare, depending on the density of planting and the area. The costs of buying, rehabilitating, and managing 200,000 ha could cost $2-4 billion. By comparison, this is equivalent to the $3 billion that the Forest Service might spent on roads in roadless areas. Larger areas would reduce management costs; smaller reserves require more intensive management and habitat manipulation.

Although the Palouse is not considered a "priority habitat," it does need protection, and it should be a part of the Biosphere Reserve Program. A large area of undisturbed grassland would be a valuable ecological baseline, for comparison with the domesticated landscapes of farming and grazing; variables, such as productivity or the effects of climate, could be compared between them. And, native species would be preserved.

The Palouse is common, meaning generally known and widely used, because it is so productive and forgiving. Being common does not mean being without value, however. Having fewer organisms or simpler patterns does not mean having less uniqueness. The Palouse has a special beauty, more subtle than a rainforest or a desert, that has become rare. If that specialness is not to disappear completely, it must be protected, now, in a reserve. Saving common lands requires a wide vision that entails making decisions without complete knowledge, ecological or economic, within the human communities that must bear the cost. Because many of the mysteries of even common habitats have not been unraveled, saving large numbers of habitats from human intervention is a hedge against the ultimate price of ignorance—extinction, whether of animals and plants or, eventually, humanity. The Palouse is a unique ecoregion, and there is very little of it left in a natural state. Let us save it.

Figure. Palouse Reserve.

Make an Ecological Development Plan: Palouse Ecoregion

This experiment is an approach to comprehensive planning based on the biohistory of an ecosystem, the cultural values of the people, and knowledge of the limits for sustainable development. This approach makes the limits explicit and sets equitable goals within the limits. A synthetic framework provides for the health of the ecological system, as well as for the health of its human inhabitants.

This plan considers the whole system and design communities for an optimal fit within the limits of the system. Ecological planning considers an optimum population within one ecosystem, although it is connected to others by trade for some necessities or luxuries. This kind of planning is a conscious adaptation of the benefits of technology to the traditional idea of physical limits. Using the Palouse ecoregion (or bioregion) in the Northwest United States as an example, we can outline a comprehensive plan to deal with some of the limits and goals. There are 5 general steps.

1. Identify our place within its natural boundaries. The Palouse is a uniquely identifiable ecosystem, with recognizable boundaries and a unique history and character.
2. Calculate the optimum amount of wilderness to preserve the natural cycles indefinitely. If the current area is less than our calculations, as in the Palouse, restore the difference and set it aside as a reserve.
3. In the remaining area, zone areas for appropriate use, including conservation, preservation, and artificial areas (with historical, cultural, and functional importance).
4. Identify the resources available for human use, including raw materials and the productivity of the areas. This productivity can be used to calculate a base line population.
5. Apply cultural modes—in style, values, and technology—to set limits on technology and population. Preserve the cultural values. Renewable resources will sustain a population longer than energy capital like oil or gas.

As part of the formulation of a plan, we have to examine the natural and cultural histories of the Palouse. We need to understand interactions in the ecosystem, as it existed before humanity, as it was lightly settled, and as it is now, dominated by humanity.

In place of a comprehensive plan this thought experiment is offered to describe an optimum human presence within ecosystem restraints; it is a deductive, synthetic, conceptual model based on data generated from research on biological productivity, the rates of resource use, and cultural valuation. It also considers minimum wilderness preservation, air and water quality, genetic minima, nonrenewable resources, appropriate technological innovation, the importance of cultural frameworks, adventure, research, beauty, uniqueness, and other intangible experiences.

This model is in harmony with strategies for sustainable ecosystems, the conservation of biological diversity, and aspects of global change. This model attempts to work out plans and policies for long-term environmental stability by paying attention to the architecture of physical and social institutions, that is, buildings as well as politics. The goal of planning is to enhance life—all life, not just human life—so the model is not restricted to one species at one time.

The human population of the Palouse can be related to land area, productivity, technology, and culture in one algebraic expression. An optimum population of 570,000 is calculated by adding the total annual agricultural productivity (in Kcal) to the total annual resources (in Kcal) of the available land area, multiplying that sum by technological and cultural modifier fractions, and dividing that by the annual per capita requirements for food and resources. Carrying capacity calculations often just consider food energy, but all needs—clothing, shelter, transportation, information generation, aesthetic satisfaction—must be included.

Resource use would be tied to population, and probably decreased. The current residents are using 10-30 times as many resources as the native people did 200 years earlier, but do not seem ten times happier.

Ultimately, our cities depend on agriculture, and agriculture depends on wilderness—for recycling, pest control, genetic diversity, soil-making, and water purification, among other things. Our cities depend on cultures for their vitality, and cultures depend on wilderness for their context and imagery. Therefore, we must preserve wilderness areas and cultural knowledge first. The size and shape of a wilderness area is also calculated, based on the characteristics of endemic plants and animals. A wilderness reserve would comprise 200,000 hectares of land; the costs would likely be three billion dollars.

Technology would be chosen to fit the location. Personal transportation, in the form of cars, trucks, and airplanes, would change. The optimum energy rate is about than half of the per capita current rate (about $1.0 \cdot 10^7$ Kcal); at the current rate, with its dependence on large quantities of imported oil, the Palouse could afford less than 300,000 people on a sustainable basis. The root problem is how to live with technology in a mature manner. We need an ecological awareness at all levels; a humane, existential ecology, where humans are part of the system and aware of it. But that may not be enough; we may have to legislate limits or induce adherence with economic incentives, if awareness and reasoning are not enough.

A number of concrete recommendations can be made to address the goals of the Palouse:

    1. Educational campaigns for public awareness of the uniqueness of

the Palouse and of the effects our lifestyles have on it,

2. The promotion of equity in opportunities and rewards for all people,

3. The offer of tax incentives to private and public corporations to redesign their paths in the community,

4. The gradual increase in preservation of wild and restored grasslands, perhaps up to 20 percent by 1999, to protect ecosystems and resources, and

5. A geographical/environmental data base for the intelligent management of the land.

The goal of planning is community success and personal happiness, based on self-reliance in food and shelter, self-sufficiency in agriculture, and self-limitation in size and desires. If human patterns were based on mature ecosystems, civilization would be far more complex; human values would allow for the welfare of humans, animals, plants, and land. We have to be wise enough to be disciplined, to leave wilderness for other beings, and yet to make good places for ourselves.

Figure. Palouse Ecoregion, SE Washington.

Common Voices Series 1986—

Why I Worry about Genetic Enhancements

I don't know about you, but I would like to be a little taller and have curly hair; I would like to be a more durable runner. Theoretically, I could be enhanced with a few genetic changes, or at least my descendants could be if I could attract a mate eventually, with my abilities to swim or pose intellectual puzzles.

New technology is constantly altering our bodies and brains. Knives and forks reduced the muscles needed to chew and consequently the number of teeth and shape of the jaw. Technology to improve our hearing, such as implants, may allow the deaf to hear. Maybe we will be able to hear bats and whales sing. Maybe we will be able to hear the states of water, from freezing to boiling. Maybe we will hear the weather or geological grinding of tectonic plates.

But, there are always losses to go with gains, and unintended consequences for every change. If people are able to hear at so many levels, maybe they will be deafened by large noises. If we try to make people more intelligent, maybe we shorten their memory. If we add connections to the brain for computer databases, maybe we will overload our critical facilities and just accept the numbers without noticing their trends or meanings. If we just go about making improvements now, making athletes faster or larger, fixing a kidney or liver, we might make ourselves into real monsters (beings who 'point the way' forward, for good or bad).

We worry that our science will usher in a transhumanism, characterized by genetic improvements to the bodies and brains of differential classes. We worry about conflicts between the natural and the enhanced, without considering the level and misery of most people without food or shelter. What an odd worry. Are transhumans going to make the inequities worse? Or will they succumb to genetic problems that we cannot anticipate? Technology requires us to redesign ourselves, even before we have enough maturity and wisdom to share things equally anyway. Maybe the consequences will be dangerous and destabilizing. Maybe the genes will be damaged or express themselves in unintended ways in a deteriorating environment.

Changing humans involves millions of genes in thousands of contexts. Improving people in a uniform holistic way will involve tracing connections backwards and forwards in space and time. It might work, but we have to be prepared for failure. That should not be a problem; science is a form of play where most experiments fail. Perhaps, the enhancements should be tried on the rich first. And, we have to think about ways to reverse changes or make them less negative.

Environmental change shifts natural selection, and we constantly change the environment by converting it to a humanized one. Is anyone sure what the current environment is selecting for? Size and the need for speed? This is niche construction and the environment again modifies the genes. Culture and technology provide new niches at a greater rate of change. These things will influence the expression of genes that we change, in ways that we cannot predict. I would be nervous about getting an enhancement that I could not predict or control, or reverse.

I trust our scientists will adhere to common sense and a formal 'Precautionary Principle' in our excitement to enhance our genes.

Why We Still Think Too Small

We are a medium-sized mammal, with relatively good vision, but limited smell and hearing. We can see details as well as whole panoramas, but we do not always recognize big patterns or long-time trends (or very small ones). We tend to think small, however, and limit our thoughts to food, tools, shelter, and sex. That had survival value at the end of the last Ice Age, but in an accelerating industrial machine, it may not be enough.

We have refined science now, so we can picture the astronomically large and the microscopically small. Our theories describe the interactions on those levels. Quantum mechanics, for instance, is a successful explanation with predictive value. Notions of particle and field are stretched so the world might be made of something else. A Holofield? The state vector describes system as a whole. Without reference to location. This undermines the defining feature of fields, that they are creating space-time as they spread out.

Despite science, we still understand things as relative to the human scale. And, we think of things as they affect only or mostly people. We spend our time being amused by television programs about humans, even their vampire or zombie states. We watch movies about people, and read books and magazines about people. Most of our professional careers revolve around human concerns, building cars, or helping the handicapped, or selling stocks in companies. Human, human, human.

Rarely, do we imagine what the world might be like for a pet cat, or the flea on her back. Even less likely do we identify with a coyote or hawk, although each of these animals have unique life-worlds laden with meaning and information. There is no chance we think about whales and krill or polyps and bacteria. We cannot imagine existence as a hurricane or planet, a star or galaxy or universe. Only science fiction writers and readers try to imagine aliens trying to solve the problems of their civilization, by justifying the invasion of earth to mate with human women. Our own planet is so overwhelmingly complex, from its benthic food chains to its weather, from its water cycles to forest ecosystems. The connections between our planet and the asteroid belt and Jupiter are difficult to grasp. How wonderful it would be to pull all that information into our brains and then have conversations about it.

What if each person on the planet chose a specific aspect or life form on the planet and became an expert on it? What if those of us dealing with cave insects met and discussed the them, then met with a group studying cave fish, and they all met with a geologist studying limestone caves? You see where this leads. More and more groups start to cross-examine each other and make links. Other groups make links to the third level of their links. Some people get married to another studying memorization in pine forest songbirds. Some groups form societies dedicated to interactions in tropical ecosystems. Then they form baseball teams and meet to play and talk about sports, video games and the weather.

The planet has many weather systems that intermix across boundaries and scales. This makes forecasting tough. We realize that similar to quantum entanglement we have *macroentanglement*. It is too difficult for brains and computers to calculate or untangle those connections. But, if we work and practice, maybe we can think larger and arrange the connections for a good balance of wild wolves and domestic cows, or humans and sharks. Gradually, we might learn to think big.

Why We Cannot Save Elephants

Tens of thousands of years after humans saw and coveted the ivory of elephants, people are still hunting elephants, until the species is threatened with extinction. Ivory is still use for medicine and art, for paperweights and billiard balls. Poachers take pictures of the dead elephants, perhaps as memories or trophies, since they sell the tusks. Conservationists take pictures and films of the dramatic living elephants to show how impressive they are, not just physically, but emotionally and socially. We recognize the emotions in them, as well as their struggle to survive our nonstop invasions of their habitats to plant fields and build houses for our communities.

It has proven to be impossible to stop the demand for elephant tusks. There is a thriving trade in these tusks. It is supported by the fact that poachers can make much more money killing and harvesting elephants than they can by farming or making artifacts or tools. Partly, this is a flaw of the economic system. When everything can be traded for money, that is, symbols of real wealth, there is no connection to the sources of wealth. Partly it is a flaw of culture, that values 'real' ivory, although the plastic and chemical forms are quite the same. And, partly, it is because people believe that they need the 'real' thing to be fertile or artistic.

We have tried ways of saving elephants: United Nations Animal program; International law; International opinion; values related to cultures; penalties for ivory trade; restrictions on hunting. Rationing might be tried. But, none of those seem to have been effective.

Regardless of our design sophistication, elephants cannot be saved if humans do not stop spreading out into their territory. Elephants cannot be saved if there are too many people making demands on them. Elephants cannot be saved if humans approach them with inappropriate images as pets, slaves or machines). The image of a machine with interchangeable parts is ridiculous and incomplete. Elephants cannot be saved if we cling to our detachment and remoteness, watching violence and misery behind a glass wall of television. Elephants cannot be saved if humans cannot acknowledge and enforce limits and rules.

Elephant saving campaigns are not really constructive in the sense of architectural or educational campaigns because they are defensive. Furthermore, what is defended is, first, in the process of change, so it can not be saved as it is; second, it will always be vulnerable; third, it is an ambiguous concept misperceived by opponents; and fourth, it is ultrahuman. What is to be saved is the potential for evolution of uninhibited development of the animal in its ecosystem. Animals do not need to be saved from natural death, a great regulator of life, but from unnecessary suffering, experimentation, and premature extinction.

We do not seem to have the will to protect elephants. Until there is a dramatic shift in perspectives and awareness, until elephants have legal or ethical standing, until dangerous historical traditions can be discarded without destroying a culture, elephants are doomed.

Why We Need Elephants for Rewilding

At end of the recent Ice Age, elephant-like mammals roamed on every continent, except Australia and Antarctica. But, they disappeared from 90 percent of their range later, with climate change and human hunting. In Africa there may have been ten million elephants, 20 times recent numbers. In India, there may have been 2 million, again many more times the remaining numbers.

African elephants eat anything, preferring tree leaves. They can destroy trees quickly, trampling seedlings and small Acacia trees. And they maintain the savanna with their dislike and exploitation of trees. By late 1800s, elephants were almost totally wiped out from the Serengeti by ivory hunters.

Much of the Serengeti was dense forest. The Masai had abandoned it because of the tsetse fly. Colonial administrators set it aside as a park. When elephants returned in 1955, the forests started to retreat and grasslands advanced. The Masai returned with their cattle. It was thought there were too many elephants. But, elephants are ecosystem engineers, reshaping their landscapes. This results in changes of species. Lizards lose out, but antelope and zebra prefer elephant savannas; and lions and other predators prefer antelope and zebra. Elephant droppings are rich in undigested plant seeds and elephant-made waterholes help other animals.

Despite adequate rainfall and good soil, elephants kept trees from encroaching into the savanna. Likely megaherbivores shaped the grassland character of much of the planet towards the end of the Ice Age. Elephants trample hard paths that other animals use. Elephants can makes caves to mine salt; the caves are later used by leopards and bats.

There are ripple effects (cascades) with animals in a food web. Leopards that get porcupine quills are more likely to hunt humans. Most involve predators. Without Elders to keep them in line and show them water holes and routes to food, young elephants can get violent and lost. Elders are necessary for animal social structure, and memory of the best paths for food or migration. Young elephants relocated without elders started killing Rhinoceros (their musth period had also extended). Until recently, were Indian elephants (and tigers) in the Western Ghats in India. Many of these elephants are captured in the wild and used for heavy moving and carrying. There has not been enough time for the systems to react to their absence and shift away from mature vegetation.

Because of their size, food intake, and social structure, elephants define the savanna, bush or forest ecosystem where they feed. They need extensive land areas to browse for food. The population of African elephants has been halved in the past 30 years as farmers take over their territory and hunters kill them for parts of their anatomy. Habitat loss and poaching are severe threats.

Territory needs to be set aside for or shared with elephants, with as much coverage of African habitats as possible, and with corridors connecting individual reserves. The population should be encourage to be at least 1.3 million, about the population in 1970, and possibly 2 million to rewild significant areas. Likely, tourist income would more than make up for farm destruction in shared lands.

Indian elephants could be doubled from about 28 thousand to 56,000; they could continue traditional movement patterns and shift vegetation, especially in southern India, where groups could be combined. Smaller numbers in southeast Asia might be increased to 20,000, but smaller populations are vulnerable to habitat conversion and might need formal parks, especially in Myanmar, to avoid further fragmentation.

We need more elephants working for a healthy planet!

Why We Cannot Stop Wrecking the Country (Any Country)

Even a president can only watch sympathetically sometimes at a circus of lawmakers that try to balance ignorance with selfishness and idiocy with good intentions. Part of the problem is the lack of responsibilities of legislators and executives themselves. Perhaps, we need a stronger oath of office for each: "First, do no harm." That means don't steal money that isn't yours, don't start wars, and don't hoard money that should be redistributed to others, through food stamps or unemployment benefits for example, to balance the luck and greed of some. We need something to limit acquisitiveness. Perhaps we could set limits before and after some elective positions, so that congressmen for instance cannot milk the job before or after as lobbyists or advisors in inflated sinecure positions.

Capitalism goes crazy with its manufactured wealth, while governments clean up the damage to people and environments. The private sphere encroaches on anything that can turn a profit, including government programs to help people recover from private sector damage. The massive surplus of profits siphoned off from corporations is hidden in bank accounts or used for grotesque architectural fantasies in parts of the world, especially in the Middle East. Perhaps the sight of skyscrapers from paper-tar shacks inspires some people to dream more, or to steal for themselves. That money is needed for social programs or to develop permanent food crops.

We just keep building shopping centers and mediocre houses under the flag of 'smart growth.' What in the hell is 'smart sprawl' or 'sustainable growth'? Are we kidding? Five-year old children know better. The elites who own the financial system and benefit from growth have no interest in conservation or social prosperity. Their goals are profits and protection of the current distribution of wealth. What's utopian about disagreeing with that or contradicting that environmental destruction is necessary for profit? That forcing inequity is necessary for economic or corporate health? That's not utopian; it's common sense!

We need equity of salaries, within limits of ten times the lowest, or maybe adjusted to desirability of the job. Decrease the take of athletes and entertainers, which everyone would like to do, and increase the salaries of police, nurses, teachers, and 'sanitation engineers,' which often have unfilled openings. We are so selfish with professional jobs that 60 percent of the planet's labor force is working in informal jobs. This is becoming a permanent, unaddressed condition. People from Mexico are flooding into the southwest USA to work to send money back to families in Mexico, just because US 'Big Ag' crushed the competition from subsistence farmers with cheap food and expensive seeds.

Scientists wait for a computer singularity to solve problems of climate change, energy generation, and hunger. Shit, we can do that with common sense. How naïve is that? They cannot solve ordinary problems, much less super-problems created by immature machines.

Tragedies happen. Bureaucratic jealousy can move mountains; the stain of an imagined affront can cost a million lives. Arrogant, but inadequately educated, governors and congresspersons can interfere with the flow of people or suspend the government until they get their way, which often revolves around a misunderstanding of government taxes and civilization.

Ordinary citizens are not much better. It used to be that men responded to an insult with a challenge to a duel, that might leave one or both injured or dead. Nowadays the insult is kept secret until the offended party can fetch a gun from the truck and ambush and murder the 'perpetrator.' Men unhappy with their third-grade experience ambush and kill students in elementary schools. To err is human. How true. But our errors are getting larger, more inclusive, and dangerous to everyone. Maybe we should redefine ourselves: To be human is to err. Forget forgiveness.

Bring Back Danger! Rewild Europe & the Continents

The whole planet has been getting poorer for the past 60,000 years. Modern humans in Africa removed several large predators as they learned to started hunting. Successful humans repeated that pattern everywhere they went, first inserting themselves in the food web as they migrated to other continents, then by hunting without restrictions. The early settlement of Europe resulted in overkill of favored animal species, causing ripples of extinctions; fire was used to drive animals or shift habitats towards grasses. The heavy human population of Europe resulted in habitat destruction, and fragmentation. Favored exotic species were brought in. Soil, water and air were polluted, contributing to hydrological and climate shifts. Larger populations, increased poverty, ignorance and greed contributed to destruction (and human misery in a positive feedback cycle). Agriculture and cities spread 'wildly' converting billions of hectares of ecosystems into agricultural fields, cities and thoroughfares between them. Now, with greater efficiency of manufacturing and intensification of cities and agriculture, millions of hectares have been abandoned—this pattern is being repeated in many nations on other continents.

Dave Foreman, Reed Noss, and Michael Soule have suggested rewilding the Great Plains in North America. This idea has a great possibilities for other continents, such as Europe, South America, Asia, and Australia. In Europe, native mammals included aurochs, bison, boar, chamois, red deer (elk), horse, ibex, moose, and reindeer (caribou). Poland has reintroduced European Bison in Białowieża Forest. Spain has some wild horses. Italy and France have ibex. Scandinavia has Sami herds of reindeer. The aurochs and other unknown animals are extinct (and possible candidates for genetic reconstruction).

These abandoned areas are 'available' for the reintroduction of wild animals who could shape the systems into a more natural configuration. Even small single hectare lots could be left for smaller mammals who are gradually making their way into cities on their own. Larger expanses could be made into more formal parks. Large mammals would be introduced from remaining stocks. Mammals from other continents, as functional equivalents, such as African rhinos to replace the extinct variety in Eastern Europe, would be added to the mix. This would require management until the system sorted out the fitness and interactions. Cattle could be run in large numbers as equivalents, until wild species could be introduced. Eventually the system would be more balanced and self-renewing (and self-managing, of course). In some areas, rewilding would begin with trees, then beavers, lynx, and wolves. In others, storks, pigeons, falcons and bats. As these areas were left by benign neglect or set into motion with sophisticated management, they could be connected by corridors, probably abandoned railway lines or decommissioned roads and highways, which would link the rewilded patches and allow migrations (and genetic flows) in the whole matrix. A balance between edge and interior in an overall wild matrix is important for large species and for interior species. Cities would become more permeable and dangerous, but that might be welcome.

No doubt working, volunteer tourists and citizen scientists would become useful in helping establish some areas. Even the more artificial efforts, which would take decades or centuries to become fully functioning ecosystems, would be valuable, and would likely attract more traditional tourists, who would contribute to the local communities. The increase in ecological value would support everything from carbon fixation to clean water (the now inappropriately but popularly called 'ecosystem services'). The ecological complexity and diversity that forms in wild systems with large populations increases the health of those systems and the neighboring artificial human systems. The shape and pattern of these renewed areas is important for their self-making and renewal.

Figure. Rewild the Ocean
Poster (Credit: Nobel).

We Need Whales to Respond to CO2 Increase & Climate Chaos

When Antarctica's great whales were nearly wiped out by the 1960s, the krill population plunged instead of swelled. Whales produced the conditions for krill to survive; their excreta supported the plankton that krill ate. Furthermore, whales excrete iron-rich plumes near the surface, where the plankton fix CO_2 in large quantities. The plumes sink downwards; the iron filters through the food chain. Other whales like Sperm whales eat iron-rich prey like squid and drive a similar cycle.

This natural cycle of predation anticipates a geoengineering suggestion for fertilizing the ocean with iron, so diatom plankton would take up the iron and carbon dioxide, bloom, and die and sink to the bottom, sequestering 1 billion dollars of carbon trading, but with geoengineering costing $2-500 per ton of CO_2.

The current 12000 sperm whales draw 50 tons of iron to the surface, J.B. MacKinnon calculates, and in a year remove 260,000 tons of carbon. When many whales die they sink to the bottom, and are fed on by scavengers in a bottom cycle, keeping the carbon out of the upper cycle for hundreds or thousands of years. There are ten more species of great whale, including right, humpback, and grey. Depending on accurate numbers they could fix possibly 30 million tons of carbon. Adding large fish like whale shark and sunfish, and predatory fish like shark, cod, and tuna, many of which also sink at death, greatly increases the amount of carbon fixed.

Prewhaling numbers were roughly 120,000 Sperm whales, which would have removed 2.4 million tons of carbon (worth $20 million MacKinnon figures). Prewhaling numbers of Blue whales were possibly over a million; they would have captured tons of carbon. Precatch numbers for the other whales and large fish are estimates, but the numbers are very high. All of these animals would be fixing carbon, and many would be returning it to the sea bottom, although the actual numbers are fuzzy, due to beachings and human takes. In carbon trading terms, a rough guess is $13 billion per year (at a much lower cost than geoengineering and a much higher return).

Obviously, we should rewild the oceans and hire the whales and fish to fix carbon. It would be cheaper, although a ban on most fishing for 20 years could cost (out of a gross of $85 billion, estimated by the Pew Group), $60 billion a year in temporarily lost income—a heroic sacrifice to be sure, but then our industrial vacuuming caused the problem. Eventually a healthy renewed ocean could easily support an estimated 1.75-2 billion people on an indefinite basis (this is a calculation for an optimum, not a maximum of 3.45 billion). It would be impossible to support the current population of 7 billion or even 3.5 billion at a minimum survival, much less a generous level of luxury. The value of the fishing in a healthy system should go above $50 billion per year, although that could be higher with more valuable fish in an ecological economic system.

It is also more ecologically sane and naturally renewable for the long-term. The aesthetic value of whales and fish is apparently much higher than the ecological value. Tourist interest translates into additional billions of dollars for community economies. Total tourism dollars for 2009-2013 were over $1.3 trillion USD (World Bank numbers); Since the bulk of human population lives within 135 kilometers of the ocean, a healthier ocean could generate conservatively over $500 billion USD per year. *We need more whales!*

Why I Don't Always Agree with Bill McKibben

Just so you understand, I am a radical liberal conservative who enjoys poking holes in the arguments of biblical proselytizers, paleolithic con-servatives, offensive writers, and ignorant news analysts. I doubt if I have ever penetrated their dense ideological hides. Sometimes, though, I worry about the people like me, who want to 'save' the planet, with good ideas, weak suggestions, inadequate goals, isolated actions, or utopian technologi-cal fixes. I do not want to discourage them, just to balance their ideas with common sense and easier cultural and ecological solutions to the civilized overkill of our success at dominating and transforming the planet.

Nature has always been changed by the interactions of living beings, so nature, and the special areas we now call wilderness, were always being modified, by Cro-Magnons, by native peoples and finally by industrial soci-eties. We never needed a name for wilderness before we were able to create the artificial ecosystems in cities and became detached from the very envi-ronment that still supports our urban adventure. But, I disagree with Bill McKibben here; there are at least eight kinds of wilderness, now, and *we do need* a wilderness designation to protect those species who cannot live with humanity at all; and, we need it to allow natural cycles to continue operat-ing and provide us with 'services' (not services really, so much as opportuni-ties to take things we need, such as fresh air and clean water).

The environment is always mutating and developing or dedeveloping. Nature never has been unchanging and static, which makes it difficult for preservation, but also for people like McKibben, who think the planet needs a new name because humans are releasing carbon. The atmosphere may be shifting due in part to industrial release of carbon, and the change is prob-ably dramatic, but it is well within the scope of earlier changes, such as the Oxygen Catastrophe or the sudden warming after the clathrate releases, so it really does not merit a new name, like *Eaarth* as McKibben offers. It may be uncomfortable to humans who expanded during the long summer, but that was not an unexpected dramatic change.

Some new organizations, such as 350.org are impressive in organization and public actions. But, I worry that we're putting our faith based on a number for carbon that is a calculated threshold. Even getting to that limit will not solve the problem of too much carbon in the atmosphere. To derail the momentum of our industrial processes, we will have to set the number much lower in the hopes that the final amount will stabilize at the (fuzzily) calculated number of 350 PPM. If we were to aim for 240, we might hope it would settle at 280 PPM, well enough below the threshold to keep the planet cooler.

These organizations and speakers are important; I have volunteered for them, and you should, too. But, we need to make them more realistic and effective in order to undo the damage from over 200 years of industrial conversion and momentum.

I also worry about the proposals for climate engineering, especially the ones that are cheap enough for a rich person or nation to trump common sense. Humans are not masters of the planet; never have been, will not be, most likely. Technology-driven processes are creating crises, like climate. Solving them with technology may create new problems. So, we need to change our mentality towards common sense and the effectiveness of radical conservation.

Additionally, recycling, green roofs and certified buildings are important things to do, but they are not enough, as they are just improvements to an industrial machine hurtling into the future—a future that is being compressed into a wreck of unplanned junk dumps with living quarters on top for unwholesome panoramic views. We need to employ traditional cultural actions to reduce population and consumption and to increase education and playful invention.

Microthought Experiments

1. Why I Don't Use the Word Anthropocene
In discussing the history of the planet, scientists use specific terms to designate eras of time, and those dominated by typical animals of those times. Thus, we have the Age of Dinosaurs or The Holocene. It is possible to recognize major eras by typical species or kingdoms. Certainly, we can recognize a cyanobacteria era or an algae era (eukaryotes era). Other eras are characterized by geological processes, such as the two volcanic eras, with their influences on the atmosphere and geological formations. The continental drift background or the extinction background. Or other meteors.

There are other contenders for a new pop science name, such as The Ecozoic Era, or Homocene Error, or The Obcene.

Now that we have identified the 'Age of Man,' since we are a dominant species, much like the dinosaurs, some, after Eugene Stoermer and Paul Crutzen, have taken to calling it The Anthropocene. This may not be the best term, although we are causing a major extinction event on the planet, on par with massive volcanoes or earth-altering meteorites.

We also want to believe that we are controlling the planet. Certainly we are having major influences, especially on species and on conversion or simplification of ecosystems. But, we have had very little luck changing the atmosphere back to where we want it. We have not been able to stop or alter large oceanic changes such as warming or acidification or even small ones like tsunamis. We are impotent to stop earthquakes. What should we call this era? How about the Holocene, which is what it was originally? The Holocene includes the brief Anthropocene as a difficult exaggeration.

2. Remove British Accents & Blondes
Could we please remove people with British or Australian accents from News programs (in the US). They don't add anything, and the sound is inappropriate for reporting serious news or tragedies. Ditto with young, toothy blonde women. Just try to showcase someone intelligent, regardless of looks, but with an acceptable Mid-west accent. Please.

3. Restrict or Eliminate Television
Many modern things can distract us: Television, television in public places, television on computers and tablets. Should we limit television to certain hours of the day or days per week? Could we emphasize Public or Educational stations? Ban reality television entirely. Limit ads to 3% of the total time, instead of 18% as it is now. Not every person in every culture wants to see a television or to be near one.

4. Equity Now!
Why not. The logic is irrefutable. If everyone was rich, minimally, then there would be less unhappiness and crime. The economy would be smoother and there would be less cheating. Come to think of it, why do we allow so few rich to have so much money? In 2005, there were 768 billionaires, with $2.6 trillion. That same year there were 7.9 million millionaires with $30.1 trillion. They could all fit in an average-size city. They seem to control 90% of the entire world economy (including hoarding).

5. Forests for Peace
Forests, violence and stability; Forests and peace. Of the fourteen nations requiring UN peace-keeping operations since 1990, twelve of them have lost over 90 percent of their forests. Perhaps peace-keeping should be abetted by tree-planting. As ecosystems are destabilized, nations become less stable.

6. *Where did all the Buffalo Hides Go?* or *Kill to Look Good!*

Where are the buffalo hides? Think about it. Probably a billion buffalo (Bison) were killed in the 1800s. Over 75 million hides were shipped to American dealers. Where are they now? In attics, as robes or hats? Lost? Degraded to dust?

What about beavers? They were almost exterminated for fashion. Where are their hides? In hats? Who of you has a beaver hat?

What about wolves, otters, or passenger pigeons? Well over 5 billion passenger pigeons were killed. Are their feathers in hats? Pillows?

Have we forgotten the tall grass prairie and the short grass prairies, with their bison, deer, and marmots? Biological diversity isn't just being eroded, it's being killed, and worn, then discarded after a fashion.

7. *Why I am a Feminist*

Carolyn Merchant and others have noted that ecocentric ethics is still a human system of thought, and its holism still masks the gender and racial differences of individuals. Of course, holism is a response in part to the rampant individualism that is threatening the environment. Venanda Shiva, an Indian environmentalist writing about the Chipko movement (tree-hugging—one of the most dramatic efforts to protect forests), argues that the masculine European paradigm (way of thinking and operating) is life-destroying and must be replaced by a life-enhancing paradigm that can emerge from the forest and the feminine principle. Perhaps a new interpretation can balance holism and individualism, masculine and feminine approaches to ecology and politics. I'm a feminist to help restore balance, not to eliminate the masculine principle.

8. *Let's Bring Back Extended Families*

I'm lonely. Sometimes I want to talk about something, an itch, an idea, whatever. But, there is no one around. Oh, there are a few neighbors, but they have different interests. The professionals are working and can't even have coffee for a break. So, I talk to myself. I used to have close friends, when we all had jobs in proximity, but then we all had to go find other jobs in other states. The telephone or Skype are useful, but not quite adequate. I want uncles and aunts, nephews and nieces. More old folks to hang around. More young to explain technology to me, as I remind them how it got that way. Now, we seem to be dispersing in our declared independence.

9. *Is Nature Experimenting with Us?*

We are experimenting with the planet, thoughtlessly and recklessly, as well with a series of conscious thought experiments, but nature is also experimenting with us.

Why I Don't Use the Word Sustainable

What do we want to sustain? Three billion hungry people? A million billionaires? Let's think bigger and better. Let's set a goal of more than mere survival at unacceptable levels of unfairness, poor distribution and prodigious waste. People have said that we should sustain natural capital, but protecting it might be better. People say we can have a maximum sustainable yield of fish or trees, in habitats in equilibrium. This is foolish; we can't keep an equilibrium and we set them up for catastrophic decline. We say forest or ocean use is sustainable, meaning there are no extrinsic costs, but our use causes the destruction of communities, disruption of processes, and pollution of air and water. If we had quantitative measures of population limits and health, they might be criteria for sustainability—but, we do not. Any truly, mathematically sustainable harvests are going to be significantly less for a long time, until the damage can be corrected by taboos (a modern taboo might be an emotional rule, such as stopping whale hunts and 'research harvests') and restoration.

Sustainability, having become vague jargon that encompasses everything, without any quantitative meaning, basically is used to advertise activities, from recycling bottles and sewage treatment to additional education, to let us feel good and to support the current regime at increasing levels of energy and conversions of ecosystems. The word is ambiguous and overused. It covers too many things out of context. Even some of the redefinitions are backwards, such as "Used properly, it [sustainability] describes practices through which the global economy can grow without creating a fatal drain on resources." No! The economy cannot grow now without precipitating a fatal drain and collapse. The economy needs to shrink, along with the population, to a measured size.

Ecological sustainability is closer to the dynamic stability of the system, which is described through constancy, accommodation, resistance, and resilience, which keep the system diverse and productive through development from a pioneer state to maturity and dissipation. The definition of this kind of sustainability is: *The ability to continuously function with optimal values of productivity and health in a human society in a developing, limited environment, using minimal resources and maximal recycling, without interfering with the basic operation of the self-organizing and self-renewing planet for the extended present and indefinite future, which requires large wild areas of reservation and conservation; the values of human society include the equitable distribution of resources, rights, and opportunities, pursued through ecological economics and politics, for an optimal population based on the limits of productivity and invention, the biological and cultural carrying capacity—balanced by ecological designs.*

Indiscriminate use has rendered 'sustainability' a worthless buzzword for selling sustainable toilet paper, although some groups, such as the US Green Building Council, have attempted to quantify and standardize criteria (so far in limited ways that include qualitative judgments based on old technology and high carbon and energy expenditures). Nothing really effective is going to happen unless the standard of living for all is balanced in food, energy, and goods, by for instance, a five-fold reduction in the US and a five-fold increase in Mali. There can be no real sustainability under current global biological and cultural circumstances! A new word is needed.

Short definitions of sustainability are continuity, durability, lastingness, and keeping in existence; certainly these are good clues to its original meaning. When we say, for instance, a building is sustainable, we mean it is *durable* under a set of standards and conditions. A new scientific word for 'sustainability' might be *ecotinicosus*, meaning a 'house of extended fitness.' The neologism may be awkward, but it sounds serious. The house has been used as a metaphor for systems of course; ecology is the study of the house. And, the word is not likely to be misused.

Why I Dislike the Term 'Ecosystem Services'

Everyone is concerned with Ecosystem Services: Are there enough of them? What do they cost? Are they unevenly distributed? Will Nature get pissed off and stop offering them, or hold us hostage to price increases? Don't worry, the ecosystem is not providing services to us. A service is the action of helping or doing work for someone. Or, it is an artificial system supplying a public need such as transport, communications, or utilities such as electricity and water. Nature is not helping us or working for us. Nature is not selling electricity or water to us. Forget it. That is not how nature operates.

The function of nature, of ecosystems such as forests, or animals like crows, or plants like trees, is to live. A tree is not trying to provide a service to anyone. It is growing to catch light and make sugar. In doing so, it may provide opportunities for food or protection to other organisms, some of which like humans may kill the tree for their own purposes. I hope humanity is not killing nature for its own purposes, like we kill trees and animals.

Organisms like trees produce wood and oxygen. They contribute to water cycles and carbon and oxygen cycles. They are regulated by population processes and limited by the physical environment. They adapt to changes in the environment, and are governed by the laws of thermodynamics. But, we need to stop pretending that trees give their lives for toilet paper in a spasm of altruistic sacrifice so humanity can eat everything it wants and wipe itself with soft tissues. They do not. We take their lives for profits and trivial needs for the most part.

Perhaps what we benefit from should be called Ecosystem Takings. And, those should be limited by the percentage that the ecosystem can tolerate and still renew itself—and only half that or less, since we do not know the actual fuzzy threshold, and since we have trouble stopping ourselves from rushing to any threshold and maybe overshooting it just because we need to grow. Overshoot is costly to the systems and dangerous for the members of the communities. Exploitation—and every living thing does it and at the proper scale—is stimulating to species, but it has to be limited to an organic scale, not an infinite or undetermined technological possibility.

Perhaps we should pay equal attention to our ecosystem givings, such as clouds of raw energy or mountains of nitrogen or seas of exotic toxic chemicals, which are often different from or in excess of what the ecosystem can accommodate or even resist. We should limit those on a system basis. We should be sure to give back large coverages of native vegetation, give back the apex predators that help balance a system by limiting the cascade under them, give back the proper chemicals such as carbon dioxide and water, and finally give our unadulterated bodies to the ground, so that the system can renew itself with the proper level of giving and taking. I hope we can figure it out in time.

Why I Never Listen to Futurists

I enjoy hearing futurists talk, but I have stopped listening. The last one I heard was waxing eloquently and in excruciating detail about the technological advances and short trends with telephones and computers; he made a number of short predictions, only a year or two—still, it was the future—about how we would benefit from using them. Oddly he never mentioned the skewed distribution of these things to rich nations. No doubt many Africans and Asians would have to share used and recycled items. He was focused on sophistication and empowerment, while ignoring the real material and social costs and damages. He never mentioned the middle and distant future or gigatrends lasting hundreds or thousands of years.

I have heard other futurists describing possible future disasters, while ignoring the present ones, the very real emergencies out of sight. They seemed to be wearing technocentric goggles that focus on technology to the exclusion of the vast physical and biological events that make the future unpredictable. They were silent about the large trends, extinctions, ecosystem collapses, and social failures. It was all just more of the same, a single short-term direction with raising all the multiple possibilities that get woven into the future of next year. So, I discount much of what the futurists could deliver of value and use to me, but I was too lazy to research the accuracy of their recent past predictions. I did remember that Arthur Clark was right about satellites, and Isaac Asimov was good with robots. But, Asimov seemed to go awry with omniscient, benign controllers. R. Buckminster Fuller should have been right with solar and alternative energy, except that he ignored the economic momentum of the vested interests in fossil fuels.

I understand that there might be a role for Futurists in the short-term business world, to help them look for longer-term trends. They might provide insights into organizational changes to deal with changes, challenges and problems. That seems to be a profitable audience.

I want to listen to presentations of thought experiments regarding the technology and trends that would extend to 50 or 500 years. I want one futurist to ask the question "What if?" to explore possible futures. Only that will allow us to anticipate the connections and outcomes of a trend that could have immense effects as it extends time. To prepare people for the surprises that appear from the interplay of complex systems as they generate the future. Futures could be categorized as possible, likely and preferable, then their actions could lead towards the goals of one of them. I want someone to look to the long future, to analyze every possible thread, to suggest how some are undesirable and some are. To some extent, change can be managed, traps and catastrophes can be avoided, but people and their cultures need to be flexible and adaptive to adjust to those changes and trends.

And, I did ask a new uncomfortable questions, but all I got was a description of a smart phone enhancement. I made a note on my smart phone and turned it off (I could check my email later). Some futurists seem to live on hope. As Ben Franklin noted, however, those who live on hope die fasting.

Allocate and Ration Water!

Fresh water on the planet is unevenly distributed; 20% of fresh water is contained in the US great Lakes, and another 20% is in Lake Baikal in Russia. The rest is in rivers, aquifers, and lakes, for the most part. River ecosystems play an important role in regional and global biogeochemical cycles, by transporting materials to the ocean; 90% of landmass is drained by rivers.

Humans have made large-scale alterations in water and material fluxes in rivers, adding additional erosion, and quantities of nitrogen and phosphorus, which can cause eutrophication. They have dammed most rivers, some rivers many times, to get water for agriculture in general and irrigation for dry areas.

Unfortunately, the health of rivers is declining across the planet. In the USA, the Colorado has not reached delta since 1984. The Trinity is wastewater. The Nile in Egypt can no longer carry silt to the Mediterranean or maintain its delta, with disastrous effects on fishing. We need to improve our management of water, especially of rivers. We need to restore rives so that they can continue contributing to seas and oceans, by carrying sediments, minerals and nutrients, which alter coastal wetlands and support fisheries.

At this point we need to start addressing the overuse and overexploitation, as well as the disturbance and interference of every ecological aquatic system—and they are all ecological—to avoid catastrophic collapses.

A hypothetical example of water *allocation* from a river follows.

Available streamflow from tributaries or rainfall: 100 gallons per unit time

System entitlements for basic ecosystem maintenance without humans: 50

Basic critical human need: 5 gallons
Basic critical livestock need: 10
Basic critical industrial need: 12
Luxury human need: 17
Luxury livestock: 17
Luxury industrial: 15

Claims on the water total 126 gallons on 100 gallon flow, obviously unworkable. Critical needs total 77; luxury needs 49 gallons. *Rationing* need only apply to luxury use; 10 could go to human, 8 to livestock, and 5 to industrial—the industrial process might substitute air or reusable oils in this example.

In the second hypothetical example, the river requires a higher water flow just to keep the system healthy.

Available streamflow: 100 gallons per unit time
Basic river ecosystem maintenance without humans: 80
Basic critical human need: 5
Basic critical cattle need: 10
Basic critical industrial need: 12
Luxury human need: 17
Luxury cattle: 17
Luxury industrial: 15

In this case some basic needs have to be rationed for industry or livestock, and the luxuries may not get filled at all. Critical needs should be filled first, regardless of cost.

A formal pricing system would help for rationing. The most common pricing policy for water, however, is a *flat rate charge*, which is set primarily to recover costs. Flat rates are not set according to the volume received, although volume is usually the original basis for the charge. In agriculture, the most frequent basis for a water service fee is the area irrigated, rather than the kind of ecosystem and local climate. For residential use in the USA and other industrial nations, flat rate charges in urban areas have been based on the number of residents, number of rooms, and the number and type of water-using fixtures or measures of property value.

Flat rates have been criticized because they do not include incentives for limiting water use or for rationing water in line with willingness to pay (not ability). Assuming adequate or plentiful water, flat rates are simple to administer and assure that the supplier gets adequate revenue. But, adequate water is a declining situation. Rivers and aquifers are being stressed and sucked dry. Aquifers especially have replaced rivers as a source of water, in the western USA, southern Iraq, and many other places.

The high cost of volume-pricing, for installing and monitoring meters to measure use, has been identified as the main reason for continuing the flat rate approach. This argument is strong in situations where water is, or was, plentiful, supply costs are low and managers are not concerned about rationing with higher prices. In other instances, water managers are using volumetric pricing to address water scarcity, which is more typical, and the high costs of developing new supplies, which are increasing. In practice, pricing does restrict water use.

Policy-makers who are concerned with allocative efficiency, that is, maximizing net social product as the goal for a pricing scheme, advocate *marginal cost pricing.* Marginal cost represents the incremental cost of supplying a good or service. It is a schedule of costs related to quantity and typically rises as further increments are supplied. When water prices are set at the marginal cost, rational consumers demand additional water only as long as their willingness to pay for their demand exceeds the incremental costs. Theoretically, marginal cost pricing produces the most economically efficient allocation of water. However, the health of the river ecosystems has to trump allocations for other purposes—this is logical, keeping it out of the price system, because no river equals no water at all for many downstream needs. Ecological budgeting and accounting demands vital allocation first.

People have gotten used to buying water, cheaply, regardless of the effect on rivers or aquifers, and they will be reluctant to change or to reduce their demands. So, it is critical that ecological allocations not be compromised. Strict allocation and/or rationing will work only when people realize water's importance, as they have in earlier emergencies or military efforts.

Figure. Woodford Creek at Mountain Grove Forest in Oregon 1999.

How Can We Protect Water?

The systems scientist Donella Meadows notes that any growing system runs into some kind of constraint eventually. In the outflow or inflow or in the stocks. The higher and faster a population or system grows, the farther and faster it can fall apart. With exponential growth of a nonrenewable resource, even a doubling or quadrupling of finds of the resource, such as water or oil, offers little added time for developing alternatives. Also, the more the growth loops evade the control loops, the worse the fall after production.

In any commons system, there is a resource commonly shared. If it is eroded, then it can become a tragedy. The tragedy of the commons arises from missing, or long-delayed, feedback from the resource to the users of the resource. Some systems are thus allowed to get worse; this is a drift to low performance, with accompanying waste. Rivers and air get dirtier, the tragedy slows, and we adapt to it. All of the ways to avoid the tragedy are cultural ways, whether through education or privatizing to regulation. Rich countries sometimes transfer technology and capital to poor, never questioning that capital or technology might not be the limiting factors. It may be clean water.

We need to learn to value water. There are three use values for water. Direct use value refers to goods that are consumed or enjoyed, such as water, food, or views. Indirect use value refers to environmental services, from ecosystems rather than individual animals or resources, such as climate or river flow. Option value has to do with the possibility or potential of getting benefits later. Nonuse values—there are three also—derive from the benefits of ecosystem services without using them in any way. Bequest value means passing on ecosystems to future generations of humans. Existence value is what people derive from knowing that something exists regardless of its potential use. A final category of value, self-value is the value of nonhuman beings and systems for themselves and their needs.

Simple conservation makes the most sense, as we just reduce the flow and have less to bury or deep-place in the ocean, or avoid with solar screens or aerosols. Efficiency alone could allow reasonable consumption. Conservation may include additional rules on the size and efficiency of industrial use, not really different than other restrictions.

We can identify nontrivial risks with water. We can identify areas of ignorance and fill a few in. But, we have to be willing to live within certain ecosystemic and biospheric processes. We also have to accept precedence of the health of the whole system. We have to accept multigenerational equality and commitment to general dynamic goals for health and fitness of the whole system. And, of course, have to deal with current inequities between groups and nations, as part of the program to fit in the planet.

Cheating is a trap; it is avoiding the rules. The solution is to re-design rules in direction of achieving purpose of rules. The trap is seeking the wrong goal. The way out is identifying goals that reflect the real wealth. Focus on result not effort. Where and when to intervene. Meadows cites Forrester's leverage points of growth, economic and population systems goals are parameters, which can be leverage points, and can make big differences. Buffers are leverage points. Buffers are large stabilizing stocks. Delay length is a high leverage point. Balancing feedback loops is a leveraging point. Governments and corporations are drawn to a price leverage point, but sometimes push in the wrong direction.

Meadows suggests that strengthening market signals, such as full-cost accounting, do not go far due to the weakening of another set of balancing feedback loops—those of democracy. The self-correcting feedback has been disrupted by (mostly ignorant) leaders who control information. Educating and designing government here might help. Taking away the power to influence elected representatives with money might be a start.

Face of Gaia in a Rearview Mirror

Many religions regarded the earth as a god or goddess, and many of other phenomena, from lightning to wind, as gods as well. As gods became more universal, the earth became an artifact made by one god. As societies became more secular, the earth became the outcome of natural processes. Yet, when modern ecology strives to think of the planet dynamically and holistically, it returns to personification. To identify a collective global 'mind' immanent in the cybernetic structure of the global system, James Lovelock used the suggestion of William Golding to call it Gaia, after the Greek earth goddess. Nature was to be recognized as active, resilient and powerful—as Gaia.

James Lovelock and Lynn Margulis put forward a hypothesis that the planet exerts a living control of the atmospheric and hydrologic processes to maintain minimum conditions for life over long periods of time. At a scientific level of thought, the Gaia hypothesis extends the ecological doctrine that all things in nature are densely subtly and systematically interrelated until it includes humanity, ethically and mentally, as well as physically. The entire earth is envisioned as a unified entity, actively shaping the material conditions of the planet for the purpose of maximizing the survival and variety of living beings.

Lovelock considers that the role of humanity in a Gaian system is first, to recycle carbon and other elements. Now, he thinks that it might be to communicate for Gaia or perhaps to assist with the controls. Lovelock has expressed concern that humanity can impoverish the whole system by reducing the total variety. Although the planet seems to be very much self-

regulating, changes in the atmosphere as a result of human transformations of ecosystems and wastes could trigger a new equilibrium that would
be devastating to human civilization. The problems have much to due
with global scale and global time lags compared to the two-year horizons
of human business. Perhaps a great problem is simply too many humans.
Perhaps a greater problem is an aging earth, too inflexible to recover from
a small fever.

Lovelock refers to the interglacial state as a fever. For life, a cooler
earth may be a safer response to solar increase, but there is not a lot of
evidence that it was more productive during Ice Ages, even having a greater
land area with vegetation and fewer deserts (or rather a large area of ice
instead). Lovelock states that there is only a small chance of reversing solar
atmospheric heating. He argues that, because of solar increase, Gaia would
have greater control during glacial epochs, which has a lower CO_2 concentration in the atmosphere, which he interprets as indicating that the biosphere was healthier and more productive, because cold ocean water is more
biologically productive. He states this without noting that the oceans are
relatively biological deserts, and life on land may be more critical for cycles.

The argument needs to be filled out, since some data of the carbon
composition in the deep ocean indicates that there was less organic carbon
being fixed. Was the CO_2 too low for plant productivity, even with a larger
land surface available near the equator? He notes that the rainforest is an
adaptation to recycle water in a warmer environment. And, it is relatively
fragile. And now it is important for carbon sequestration as well. The ocean
deposition of CO_2 is important of course, as a physical process of the dissolution of silicate rocks, and as the biological flow towards a sink.

Think about this: Ice caps cool the atmosphere and lower the sea level.
More land is exposed in the equatorial belt, which absorbs more heat. Do
trees make the difference, creating more clouds? Are cool ocean currents less
cool in glacial conditions? Do they bring up more sediments or less? Is there
less sea life than before? Was CO_2 too low for more productivity.

Perhaps the problem is one of cognitive dissonance. We humans have
impressed ourselves with technological sophistication at a local level. But,
we often ignore the effect of scale. Before dissonance can be resolved, the
scale has to be right. Too big and the exchange is ignored; too small and it
is ignored. It has to be a proper human scale, a proper gift exchange, or just
what is understood from personal experience. Otherwise, self-deception or
inattentiveness is the result. Personalization of large systems or 'entities' can
help us understand the network of exchanges.

Evolution eliminates many options and pathways. Natural events destroy billions of living beings. Is Gaia cruel, therefore, like Kali or Nemesis?
Or, is that a problem with personalization? Earlier, Lovelock suggested that
metaphors were crude ways of knowing, but later he emphasizes that they
are needed, however, to comprehend the earth. Darwin had described evolu-

tion as wasteful, blundering and cruel. But, cruelty requires consciousness. And, we humans consciously drive plants and animals extinct in a cruel way. Gaia only filters. Gaia is not a cozy mother and cannot be propitiated by gestures like carbon trading or sustainable efforts. The earth can be benign like ancient goddesses, but also ruthless. We are not separate from Gaia.

It is not likely that we will have to save the earth; it can save itself, as it has done before. But, we may need to save the environments that we like. We have to save environments to save the charismatic animals and plants that we love to observe. And this is what ecological design can do. Gaia needs ecosystems on land and water for self-regulation. And, this is what global ecological design can ensure. Although the Gaia hypothesis renews the idea that the earth is a mother for us all, and reinforces our under-standing of the interconnectedness of biological processes, it falls short of demanding our responsibility and relies too much on our consciousness, which is usually focused on our hands and feet, and on fast food and fast vehicles. We seem to be speeding away into the technofuture too fast to do more than look behind us and shrug. We need to become responsible.

Why the Earth is More like Gaia and Less like Medea

Many people have complained that the planet actively threatens human civilization with tectonic changes and extreme climate events. In his *Medea Hypothesis*, Peter Ward claims that life is predatory and self-destructive and will return the planet to its abiotic condition. The hypothesis raises some interesting questions, many of them addressed by Lovelock and others in connection with the Gaia Hypothesis, but then it sidesteps measured responses with fallacies, errors, contradictions, assumptions, and overwhelmed models.

One example is the Oxygen Catastrophe (or Great Oxygenation Event) 2.4 billion years ago, which wiped out many species of anaerobic bacteria; it also reduced carbon dioxide in the atmosphere and may have triggered the Huronian Glaciation ('Snowball Earth'). Ward argues that this catastrophe was caused by cyanobacteria and that those life forms were destroying life at a large scale by their innate desire to convert all resources.

This argument depends on the Fallacy of Simplicity, where multiple interlocking factors are ignored in favor of a single cause. The photosynthe-sis of bacteria was producing oxygen, before and after the catastrophe (and it was a catastrophe for multitudes and generations of anaerobic bacteria), but earlier some organisms and massive quantities of dissolved iron were capturing free iron (leading to 'Rustball Earth' first). Only after miner-als were saturated did excess free oxygen accumulate in the atmosphere. Before the catastrophe, submarine volcanoes reduced oxygen and removed it from the atmosphere; by the Archaean/Proterozoic boundary about 2.5 billion years ago, continental volcanoes started erupting and oxygen levels in the atmosphere increased. The evolution of plankton put oxygen into the oceans. Although oxygen poisoned anaerobic bacteria in the atmosphere and its interface with the ocean, through photosynthesis it provided much more energy to organisms; it promoted more diversity in organisms and it increased the diversity of minerals through oxidation, which provided more resources to living organisms (eventually leading to 'Slimeball Earth'). The interaction of oxygen with ultraviolet radiation produced an ozone layer that protected organisms from radiation and permitted the colonization of land. The catastrophe did produce an opportunity for the later expansion and diversification of life.

Of the five great extinction events and the five lesser, Ward blames all but one—the bolide impact 65 million years ago—on the suicidal tendencies of life. Under closer examination, all the greater and lesser extinctions occurred under complex conditions with multiple factors, including bolide impacts, volcanism (and flood basalt events), sea-level drop, deep ocean anoxia, glaciation, sustained warming or cooling, and abrupt climate change. Climate is driven by many factors, such as orbit, wobble, plate tectonics, and atmospheric gases (mostly uncontrolled by life).

Life has become a participant in climatological and geological change, so biogenic factors are important contributors, although astrophysical and geological processes seem to be more potent drivers. The crust of the planet is formed by volcanoes and earthquakes, and by the actions of bacteria, lichens and trees, which break down rock with acids and roots. Processes of life concentrate resources. Metals and gases toxic to life are isolated and end up sequestered in geological formations, although geological processes can release some of them suddenly, resulting in methane eruptions or mercury poisonings, and related extinctions.

Ecosystem stress is a possible cause of extinctions, especially compounded by a sudden shock to the system. Some species with high turnover rates may be vulnerable. Predation can cause extinction under some circumstances, when the predator has multiple prey species but concentrates on one favorite—otherwise predator and prey dance around limits. By now it should be obvious that the crucial factors are biogeochemical.

Graphics showing carbon dioxide and extinction events indicate that the extinctions are related to changes in carbon dioxide and oxygen and to changes in temperature and moisture. Obviously those are related to living processes that moderate them. The argument seems to be: 'Over 99% of all species are extinct; over 99.99% of individuals are dead. Life killed them. Death to life!' Peter Ward might say.

Ward defines species success in terms of numbers, biomass, and range, as well as ability to survive. He uses this measure of success as a way to compare hypotheses. For instance, he calculates that the total biomass peaked a billion years ago (in the Achaean and Proterozoic periods), before the Cambrian explosion, and has been declining ever since; he relates this to the gigatrend of declining carbon dioxide in the atmosphere. Ward concludes that future changes, despite increasing CO_2, will cause the death of plant life, then oxygen starvation of animals and death of all ecosystems.

Ward seems to blame the planet for not adequately protecting living systems from external events or from the actions of each other. The planet is a stochastic system. Life is an emergent process for producing experience within new kinds of order. Life changed the order of the planet, which allowed living forms to exploit various forms of matter, energy, and structures, including other living forms. Life creates ecosystems of aggregates. These systems develop so that different forms complement each other functionally. Living systems become more resilient than nonliving systems, modifying all systems while maintaining stable states. The variety of interactions increases; competition, emphasized by Ward and others, does enhance or eliminate forms, but kinds of predation and cooperation shape living systems in a living environment. The living biomass produces waste biomass that enters soil as well as geological levels. Evolution or atmospheric regulation are just descriptions of the historical development of life in the planet.

If life was so determinedly suicidal, how did it survive over a million years, much less over three billion? If it *is* suicidal, then the evolutionary evidence suggests that it is ineffective. Will life destroy the living planet before a physical catastrophe can? Will we? Is this a problem innate in conscious living—the desire for nothingness? The problem stems from a variety of conditions, from constant entropy to galactic dust lanes. The living system of the earth has to react to a variety of external events, from energy bursts to asteroids, orbital changes, solar increase, and tilts. And, these changes cause challenges, such as higher or lower temperatures over periods to which living systems have to adjust.

Life is not a loving motherly goddess (Gaia) or a murderous mythical one (Medea). Life is not benevolent or destructive, just opportunistic. Living systems do not intend to destroy every living system. They adjust to changing conditions and exploit them to stay alive. Nevertheless, Lovelock, Ward, and others have raised interesting questions. There are positive and negative feedback cycles in the biosphere. Living organisms contribute to shaping all the planetary spheres. Species do tend to reproduce beyond the capacity of the environment (and its resources), especially in the absence of other species and environmental restraints, which can end in catastrophic decline and extinction. But, does a species even measure the environment to calculate limits? Is there a law of growth in life?

The living environment is a wicked problem for design, especially now, when anthropogenic changes are contributing to the interglacial global warming. Traditional or ecological design may have difficulty with so large a system. But, global ecological design can address the difficulties at the appropriate level. Global ecological design is as subversive as any art or design; it can overturn wicked design problems. Global ecological design is a wild way of thinking that can mesh its approach with the wild planet, using an ecological perspective, systems understanding, participation, and standards of knowledge.

Can We Stop War? Why War Will Not Stop

Politics is the management of people through the equitable distribution of resources, as well as the management of relations with neighbors or trading partners, using negotiation or force—war if necessary. As most cultures have grown from bands to states, the reasons for war have gone from insults and personal conflicts having to do with bride exchange, broken rituals and personal honor, to state and external reasons, such as territory, resources and patriotism. Wars have changed in scale. From disagreements and limited conflicts, wars have become a centralized state function to gain territory of resources, often involving entire regions. The style of warfare has changed dramatically over 500 years, from general's and professional soldier's wars to chemists, physicists, mathematicians, and engineer's wars, using gases, nukes, guided missiles, and biocides.

Wars, like the recent one in Iraq, are based on weak assumptions: That the war can be waged by blasting away any threats, and that it can be contained by using conventional weapons. But, like most actions, it has affects that can get magnified in the larger system. Furthermore, war breaks out of the barriers that the participants try to create, destroying properties and civilians—there are no longer any safe buildings or noncombatants—and ruining the social and ecological fabrics.

All wars now are ecological wars that can destroy the basis of civilization. There are no nonmonotonic effects. There are no side-effects. There are only big packages of effects and they can all be measured. Perhaps the next war will be to directly destroy the ecological basis of a community, culture or nation. These wars could destroy entire generations and traditional social structures. Perhaps, future wars will have less to do with honor and territory and more to do with crises, such as famine, population, and environmental collapse. Perhaps, wars will have to do with symbols and religion, again, as they have in the past. War is now waged against cultural problems, such as poverty, and against the environment itself, as we declare there is no room for weeds or pests, without understanding the ecology of either.

War has advantages and disadvantages. One of the advantages of war is its long tradition, being simpler to understand than rights or macroeconomics. War is cheaper than before. Of course, war is also stimulating and challenging, at least for the survivors, who are bonded by the danger. Any one disadvantage is usually greater than the sum of all advantages. It used to be that it cost more to attack than to defend, with preparations and supply lines. Now, it costs more to defend against a bullet than to attack with one. Cheap bullets destabilized the western US, but law enforcement with bullets restabilized it. Possibly this will happen with cheap cruise missiles. Maybe not, as the cost of attacking will always continue to be less than defending. This means that the world may become more violent and less stable, especially at local levels. Information attacks may be even easier. Ecosystems in disputed and ravaged territories are always disrupted, damaged or destroyed. Human suffering is always made worse than it was. And, war never solves the original problem.

The historical rhythm between war and peace inevitably leads back to peace. When does peace occur, when it is won? Unlike many forms of war, peace is a process with a less rigid beginning or end. It can never be won or kept permanently. It does not have the prestige or honor of war, and perhaps this is why so much less time and effort is devoted to peace. Is there a way to limit war, within the context of peace? Is there another way to humiliate or embarrass an enemy, or let some other kind of balance be found? Is there a way to limit violence to heroes or to leaders, as if either would agree to have their individual expertise be responsible for a whole population? Imagine the size of the television audience if two leaders were to square off with cavalry swords or boxing gloves. Imagine the commercials!

Why Terrorism Continues & What Should Change

Terrorism is the threat and accomplishment of violence by small cells of people targeting other groups, usually within a different nation, using sophisticated armaments and surprise to inflict serious damage and fear. Terrorist acts have killed thousands of people in many nations, usually for political reasons. Since the 1960s and 1970s, they have targeted airplanes and groups of uninformed civilians. Many share the rhetoric of struggle and the desire for publicity. Although they have rarely had a global impact or brought about improved, lasting social change. Many terrorists are religiously motivated and supported by some religious leaders.

It is very difficult to anticipate and eliminate terrorist threats, especially random suicide bombings. Winning a war on terror is not possible. But, the idea now is deeply rooted in modern cultures. It is perceived as high risk. There are many possible avenues of attack, including water supplies, foods, power grids, buses, and skyscrapers, most of which cannot be defended adequately, without the creation of a prison state. There is such an array of guns, poisons, gases, bombs, microbe pathogens, computer viruses, and small nukes that could be used. Apparently, though,

95% of all terrorist attacks killed none or few people and were no worse than automobile accidents.

The 'perfect' terrorist attack, by contrast, would have immense shock value, likely with thousands of fatalities, broadcast live to the planet, along with a video explanation of who did it for what reason. Perhaps that possibility is why people inordinately fear terrorism and are willing to spend billions preparing for and preventing it. However, it is almost impossible to deter most of the attacks since they are ideologically or religiously motivated. And, if the perpetrator wants suicidal martyrdom, only an unsympathetic angelic visit would be effective. Military actions targeting groups may ring up big casualties and make some people feel better, but they would likely reinforce the reasons for terrorists.

So, what can be done to reduce the number of attacks, given the reasons for them? Religious conversion? Payment of large sums to other nations or groups? Public political confessions and apologies? Probably not.

But, there are some actions that may gradually reduce the need for terrorism. First, have all people show respect for the religions and cultures of others; reduce or censor the number of stupid burnings, writings and pronouncements. Next, and this will be difficult and slow, try to correct the vast and painful historical inequities between nations, classes and individuals; this would involve many simultaneous actions, such as paying living wages, taxing incomes over a million dollars at a high rate, perhaps at 90%, and investing in all the public infrastructure, from jobs and food to parks and libraries (in every nation). Work to arrange resolutions to long-standing ethnic or political feuds that waste so much time and life. Finally have nations and groups help one another, starting with international volunteer programs run through the UN and continuing through organizations such as a universal Peace Corps or Red Cross. Once people communicate, many stereotypes and hatreds may evaporate. But even then, some people will always be unhappy and have access to weapons.

Why I Like to Pay Taxes

I like to pay taxes because I like having schools, libraries and public television to learn things. I like having public parks for relaxing or playing baseball. I like having local, state and national parks for staying in more natural surroundings and observing wild animals. I like having personal security where I live (even if it is in the remote mountains of Idaho). I like having roads, with speed limits and stoplights, that lead to hospitals and grocery stores. I like many aspects of civilization, and I doubt we would have them without taxes.

I pay an income tax of about 15% on my salary of $25,000 a year, after business and charitable deductions. I also pay sales taxes on most things, like telephones, computers, shoes, and roller skates.

Traditional taxes most of us pay include Title, Severance, Sales, and Event taxes. I sometimes pay some of those. I would certainly be willing to pay new taxes if there were a tax shift to 'taxing bads' (which was introduced by Arthur C. Pigou about 100 years ago). Bads, of course, include pollution, land conversion, and mineral depletion. Taxes could be ways of internalizing costs for those bads. These taxes could include Use Taxes, which are designed to reduce the scale of output of pollution for instance. Or Element taxes, for carbon, nitrogen, sulfur and others. Or Loss taxes, for fossil fuels, gasoline—which is dramatically undervalued, it should be at least $8 a gallon—land conversion, forests, and others that make up the natural capital base of any community. We should all pay Adjustment taxes for pollution output, cigarettes, alcohol, drugs, financial speculation, and heroic income and possessions. Distribution taxes could correct financial imbalances. I would be willing to pay fees for voting, media, business, marriage, reproduction (as discussed by Jacob Javits & Robert Packwood), driving, wildlife collection, weapons (Licenses), roads, and for tourism (fees).

Of course I pay Social Security tax, since I do not have a retirement plan or account (I work for my own imaginative ecological design company). But, I hear that some of the Wall Street predators (a basketball team?) want Social Security tax to flow into private investment accounts, where they would provide the advice and services for a cost that the government does for free. Everyone knows that Social Security is a good deal. Rather than raise that tax for most of us, the government could reestablish a special tax of 90% on income over $2 million (as it was in the 1950s). Or 90% tax on estates over $5 million. Let's tax financial transactions, especially those held less than a month (especially those less than ten milliseconds).

Lets not tax incomes at all. And, none on food or homes. Extend Social Security to everyone at birth. That sounds good, and it should even let us keep more income, with all the new taxes on 'bads,' as long as we can be frugal and less wasteful. We could still live in comfort, with luxuries and more time to enjoy them.

I would pay a tax to help establish a Northern Hemisphere Wolf Path linking most of continents where wolf territories could expand.

What to Put in Women: Education & Opportunity

Men have most always thought they knew how to treat women or what to put in them. The Catholic Church continues to ban sex education, family planning and contraception. In Libya, some men marry to harvest children; life's purpose is leaving heirs, but we can moderate the numbers, since we are over 7 billion people. In the mod-1970s, Sanjay Gandhi's Indian government promoted the emergency, forcible sterilization of 8 million women and men, as a population control measure. That back-fired.

Other countries, especially in Europe, regardless of official policies towards women, and most of them were very liberal and progressive, saw their birth rates go below the rates for replacement of the population. This was especially true in Italy and Germany.

Some countries, such as India, loosened any kind of rules. India's growth rate is approximately 17.64%. Other countries, such as China, instituted strict rules. China's model for one-child per couple became official policy in 1980. But, the model missed cultural values: Value of farm child versus urban child, or son vs. daughter, or farmer vs. merchant. Was it too drastic? Despite problems with sex selection, abortions and extra children, the policy did lower China's rate to 9.53% (2009).

At that time families in Iran were encouraged to have as many children as possible, for the glory of Islam. Somewhat later, in 1988, reports of economics and budget officers were given to Khomeini, showing that soon the country could not feed its rapidly growing population. Khomeini replied, 'do what was necessary' to have country educate and feed itself. A family planning program was launched, but it did not forbid people from having more than 2 or 10. Instead, a corps of doctors and social workers visited every small village by donkey, camel, or truck. There was education in every village, especially female education. All couples attended premarital classes. People were told that 'one is good, two is enough.' But, they could have as many as they wanted. By 2012, the birth rate was below China, India or western Europe, at 1.24%. Women went to college and started their own businesses. Unfortunately, the new Ayatollah may reverse these liberal policies.

As a result of looking at the numbers and effectiveness of some programs, it turns out the best thing for women is education, followed by empowerment, with equal rights. These factors give women the confidence to behave like men, at least professionally and politically. They start more businesses and become involved in more movements to protect the environment or worker's rights and pay. This is as it should be. But, more programs in education and opportunity are needed in many countries, from Mali and Macedonia to Vietnam and the United States.

Why is Technology Such a Problem?

New Mars rovers, moon rovers, self-driving cars, exoskeletons, fast jets, 3d printers for organs, microdrones, smart watches—these are exciting products of our technology.

Due to the increasingly widespread use of ever more complex technologies and the frequently unintended consequences, problems may arise in their use that are unrecognized or only partially addressed. Technological innovation, combined with accelerating population growth, leads to clearing of many forests for agriculture. Technology also promotes land degradation. Plowing and improper conversion cause erosion.

The combination of cheap goods and complex tasks can lead to sweatshop slavery and unsolved wastes. Some problems are solved, but new ones are created by unconsidered use—problems such as toxic waste or radioactive waste. Time and leisure are needed for technological innovations. People stressed or starving rarely invent their salvations.

Technology has reduced the globe to a single, closed system, which humans can share according to their financial resources. Our direct experience of the world has become shallow, in spite of faster travel. Travel used to broaden the mind, but now it narrows it. We travel in sealed corridors like boxed goods, comforted by homogenized foods and several 'world' primary languages. Technology has distanced human experience from the meaningful time and extent of experiences. Technology or social structure can mask the internal stress from fast economic growth. Technology has made the suffering of many domestic animals invisible to consumers of animal products.

Under what conditions can the effects of technology not be contained? Does technology always escape and produce its effects on the nonhuman environment as well as on other cultures? If so, then technology is always ecological, that is, it is always part of the environment. It always generates some change in the environment that it is part of. It is not the same environment after the introduction of the extension of technology. Any addition (or subtraction) is a change, and any change has many effects on everything. The conditions of survival for all change.

Technologies change institutions and the relationships between them. As an institution adapts a technology, its view changes. A new technology threatens other institutions, which have competing technologies, as the whole mass has considerable momentum and investment. As the nature of institutions changes, the nature of communities, cities and cultures, changes. Thus, technologies change the form of tools and whatever they shape, then change the quality of response to those changes. And technology seems to dictate that what is possible should always be done.

Human activities on global scale cause mass extinction of species. In another kind of loss, our evolving social order of relationships and communities is being disrupted by powerful inhuman forces of modern technology and economics. The changes are too rapid and extensive for social evolution to work property.

Think. Culture is too fast for biological evolution, and ***now technology is too fast for culture***. What will happen? Rather than speed up culture any more, we need to slow down technology (and cultural change, to some extent). This can be done with principles, such as The Cautionary Principle, and with procedures, such as complete testing and assessment of as many possible impacts as we can imagine. Technology may be less of a problem.

Can Technology Fix Global Problems?

The technological fixes, from the modern industrial response of geoengineering, are problematic themselves. Geoengineering is the deliberate manipulation of the global metabolism to correct or adjust human interferences. This approach offers several options: Orbit mirrors to deflect sunlight, launch sulfate particles into the atmosphere, or fertilize the ocean with iron. It would consider burying CO_2 gas or solids in the ocean or land. They are simple symptomatic responses that ignore the system. If the problem is simple, pollution or fossil fuels, and the fix is simple, then it neglects the complex issues of human conversion and scale. It ignores the proper relation of humanity to the planet. Technofixes are based on an inappropriate logic, linear and cause-effect, instead of nonlinearity and multiple effects.

Americans and Russians made proposals in the 1950s and 1960s to control the weather and modify the environment, for instance, Mikhail Budyko's proposal for artificial volcanoes in 1977. How do we remove the 100 ppm excess CO_2? Atmospheric scrubbing on a heroic scale? Planting more trees and restoring forest coverage to Pleistocene levels?

Although several national space programs have been able to keep people on ships and space stations for limited amounts of time, with bottled oxygen, scientists in Arizona spent $200 million to prove that we did not have the technology to keep eight people in oxygen for more than a week, using a balanced ecosystem approach. So far, our best technologies cannot substitute for natural processes, especially in terms of detoxifying wastes.

One bad assumption about computer intelligence and manufacturing is that people and machines perform optimally at all times; *they do not*, in the case of humans sometimes, intentionally. Another is that we have anticipated all the consequences of new technologies—*we never have*. We cannot depend on computers to fix complex global problems.

Scale is problematic with global technology. What works well at one level may not adjust well when scaled up. Scale changes the interaction. If the technology is too large, there are too many interactions to monitor or understand. Scale also changes the relation to the individual. The problem of using metaphors between scales is that there are too many emergent properties that do not appear in the smaller, and better known, scale. Foam could be a metaphor between scales. Foam can be found at the smallest sizes imaginable smaller than subatomic particles, and at the largest scales imaginable of the entire universe. But, its predictive value might not be adequate. Seafoam, which is white, refracts a large fraction of sunlight than water. The interaction of light and foam has an effect on temperature and climate. Global cycles of local exchanges link scales. But the larger the scale, the longer the lag time before any changes show.

Global cycles of local exchanges link scales. But the larger the scale the longer the lag time before any changes show in interactions. All interactions are dynamic interactions between human societies and their environments, from the perspective of long-term patterns and processes. The interactions become tightly coupled. Too tight a connection leads to rigidity, which increases vulnerability to change and reduces flexibility in the system. Globalization suppresses many diversities, which are essential for stability.

William McNeill suggests that escalating vulnerability is the price of local and short-term mastery, especially when the scale is increased to the global. Joseph Tainter argues that societies choose to increase their complexity to solve immediate problems, but that leads to increased vulnerability in terms of returns on investments of energy in the complexity itself. Human complexity, like ecosystem complexity, requires demands for more energy. Globalization is a titanic effort to put us all in the same boat, more complex and advanced, but just as vulnerable to human error and natural catastrophes, and ultimately to a global collapse.

Designing Hybrid Technology Systems with Life

Given the limits of the Precautionary Principle, as well as other principles and standards, it seems possible to create living cells that could create machines. It seems possible to produce nanotools that could scrub or repair living cells. In fact, it might possible to create entire communities of machines and living beings on a very small scale that would have a vast number of interconnections to wild ecosystems. It is quite likely that these systems might be more fragile than natural systems, collapsing or dying out in months or years. And, it is quite possible that the remaining waste might not be completely biodegradable. Ecosystems already get mixed with plastics and pollutants. Given the possible dangers, however, these possible changes and applications need to be kept isolated for extensive testing. In fact, the proposals need to be critically reviewed by panels of ecologists, engineers, and other scientists and literate reviewers. This kind of explosively reproductive technology has to be planned and engineered through complex thought experiments, especially before it becomes part of the mix. Thought experiments let us describe interactions, such that if one thing and another interact and a third thing happens.

Mixing is part of the process of ecosystems, as animals and plants try to fit and excel at living and reproduction. Human and wild communities are already mixed. Humans use tools, so that means that technology and nature are also mixed; many other mammals have been observed to use tools. Humans have always collected food from mixed systems, without landscape conversion. A maturing agriculture might recapture this strategy. Mixed crops are easier to grow; we know that from experimentation.

We need to have mixed communities. This is what evolution and wildness do—provide the turbulence and chaos that every system needs to be mixed and renewed. Nature, the formal matrix for wildness and chaos, pursues a strategy of overproduction and waste. This is not the wise, motherly Nature of our dreams, this is a wild creator, profligate with energy and forms of information through low-efficiency and the waste of cells, seeds, and sperm, and with the escalating warfare of plants and animals. The sheer magnitude of energy and diversity allows tremendous margins of error. Error pushes the whole process, allowing possibilities and diversities. Flawless transmission would have stayed at bacterial level. Freeman Dyson suggests that the reign of genes, like that of the Hapsburgs is "despotism tempered by sloppiness." The harmony and stability of nature are perceived by human minds as they make them abstract.

We have metaphors to support the idea of mixed systems, including a process perspective, field concept, self-organization, and co-constrained construction. These metaphors form the basis of a new image for humanity, where we are an integral part of food webs and part of an organic cycle of birth and death. Human nature does not find meaning in an absurd world, but discovers its structure by interacting with the surrounding order.

We have learned from Charles Darwin and others that nature chooses the 'fit,' those who adapt to a specific environment. But, that nature seemed more stable and predictable than it does now. We are learning that the 'flexible,' those who can adapt to several or changing environments, may survive better under changing conditions. And, now that nature and culture are coevolving, it may be the 'creative,' who can change the conditions slightly and limit the changes wisely, who may survive even better! For instance, the North American species that were adapted to glacial-era forests died out 12,000 years ago. Flexible Generalists like humans did not.

Cultural development also changes through a process of hybridization. Industrial culture becomes a hybrid as it tries to satisfy the needs and wants of a growing population, especially for space and food. We have hybrid political systems, such as democratic socialism. We have to have goals and define new niches in mature hybrid ecosystems. We may not need as much technology. Symbiosis can be a model for cooperative designs.

Changes in growth and dominance have pushed us beyond our traditional cultural experiences. The dangers from this are immense: Slow catastrophes, such as extinctions and total conversions; sudden surprises, such as dead zones in the ocean or holes in the ozone layer as a result of overfertilizing, or overenergizing or overpoisoning; human ignorance of change and uncertainty; and inadequate responses to long-term changes.

The novelty and complexity of interactions of innovations can lead to thresholds that are tipping points, triggering punctuations in natural functions with consequences that may surprise humans. The current global environmental condition has no comparable analogs to any other previous time. One source of surprises seems to be the interaction of climate and environmental change combined with economic development and human and animal health. As tropical forests are converted to farmland people encounter diseases without having evolved resistance. We have to search for discontinuities and surprises that we may not be able to predict, although we can learn to be recognize early warning signs.

Our growth, pollution and terraforming has put us in a trap and only design and engineering, combined with severe conservation and changes in rich lifestyles, can get us out. Urbanization might help. So could technology. The new technologies promise help, but also will create larger problems. These are nuclear power, biotechnology, geoengineering, and information technology, which might be effectively multiplied by convergence, and either enhance or wreck human performance and a proper perspective for rethinking science.

We feel it necessary and important to embrace new technology. Environmentalists consider that sustainability is a sufficient basis for technological design. This may not be so. It may be necessary, but not sufficient. Perhaps some kinds of large-scale thought, such as Messianic thinking, have influenced our disposition to large-scale technological fixes, as well as the idea of redemption from the practice of irresponsibility.

As part of restoration on a large scale we have to consider whether it would be viable to green North Africa with technologies promote rain and build soil. Would it be viable to move icebergs to water the coasts of South America or Australia? What about micromachines and the use of nanotechnological devices in human bodies? We always think that technological problems are easier to solve than social problems. They are not, partly because they become social and environmental, and need fixing.

For climate change, one technological option might be large-scale planetary geoengineering, which requires equally large oversight. Energy use is so large that it requires a government. Extreme technologies are so large that they require a global government. The scale is large, as is speed and scope. Most meaningful actions have to be at the largest scale possible. But to be good at global, we have to think at the level of the solar system.

We have to let the dream of control fade, now, in order to creatively adapt to and interpenetrate the uncontrollable uncertainty of a new climate. If we make mistakes, they can be repaired if small enough. For restoration, we can experiment, improvise and innovate at a small-scale. We need appropriate technology, but we also have to behave appropriately with other forms and scales. That is, we need to keep an appropriate level of complexity. Our civilization as it is may be too complex to support.

In any discussion about solutions, we have to consider personal solutions, technological solutions and social solutions. Stopping technology all together might lead to disaster, unless we combined it with radical reductions and conservation, which may be considered in any mix of strategies. But, we can also assess our current technologies and combine them with other approaches to create a truly mixed design on regional levels.

Living in a Living Internet

The Internet is a network based on a networking infrastructure connecting millions of computers together on every continent, such that they can communicate directly using the world wide web to exchange information. Computers are nodes or connection points, capable of sending and receiving information, that connect through links (usually 3 to 5 for any connection between computers—the same as path length). A link connects 2 or more nodes physically in a network, which is composed of nodes and links. The web working on the global Internet is useful to people for communicating personally or professionally; it has transformed the way people interact.

As original as it is, the Internet was preceded by billions of years by ecosystems, which also, analytically speaking, use links and nodes for communication. Networks characterize ecological and biological systems. Like the computer Internet, each node has a degree of connectivity to other nodes. A cell can be organized as a network, as can an ecosystem or landscape (although the terms change, e.g., nodes=patches and links=corridors, and the patches and corridors are spatial units).

Organisms exist in networks of dependence. Douglas-fir trees require fungus growing on their roots. Termites require microorganisms in their gut to digest wood. Acacia trees depend on ants to defend it from giraffes, and the ants need the tree for sugars. Eating is what characterizes much of the communication; energy flows between nodes. Some parts of this ecological Internet are called a food web; a smaller part is a food chain. Food chains can extend beyond the local ecosystem to the regional and global systems, especially in the case of some birds (e.g., Arctic tern) and fish (e.g., shark).

All of their lives depend on the integrity of the network. All organisms are connected by global cycles of oxygen, carbon and other elements, which reach into soils, the atmosphere and the ocean. Some networks can become complex. Networks of eating can influence the pattern of the links and nodes. For instance, humans like cattle for food (although smallpox developed from cowpox and infected and killed many people). They reintroduced them to the African savanna, but cattle required a lot of water and compacted the soil getting between plants and waterholes. Furthermore, as their moist droppings dried, they heated up, killing grasses and forming impenetrable 'pavements' over the soil. Then, cattle started dying as tsetse flies infected them with a parasite. So, antelope, giraffes and many other herbivores returned, eating a larger variety of grasses, leaves and shoots, and leaving dry fecal pellets behind to be eaten by decomposers, which returned nutrients to the soil. Lions, hyenas and other carnivores preyed on the many plant-eaters, permitting a large diversity of plants and animals, many new nodes with more links.

The global ecological Internet is kept in place by various mechanisms, such as the transport of small organisms by wind or water. If a habitat is destroyed by a fire, the web can be reestablished over time. Humanity often increases diversity at first, but over time, simplifies the system. Leaving or enhancing diversity keeps the nodes of the network in place and the network itself healthy. Many kinds of technology, from meters to satellites, can be used to monitor diversity and in fact can make a hybrid system that can tolerate measured human exploitation (i.e., we can take food and materials, and clean air and water) and still remain healthy.

Why I am Afraid of a Global Anything

The globalization of economics, at least this implementation, leaves many things out of the equation: Local decisions, importance of representation of labor, the fragility of local ecosystems—any brakes on the engine of growth. Where is the dark matter of society? The social masses that are being affected the most, and usually negatively or harmfully? These people should be campaigning for protection of their rights, values, customs, worth, and the living environment.

Their voices used to be heard, because they had enough power, under enlightened laws, to halt corporate unfairness in its tracks. Herman Daly points out that the agreements between labor and capital were reached through elections, strikes and lockouts, not theory. The health of the local economy depends on local agreements, but now those can be repudiated by global agreements on global integration. That is a poor trade for workers, in the name of a global free trade. And, it is free of significant protections for the unique values of the participating cultures and the ability of workers to strike balanced needs. Violence in marches or strikes is treated as a criminal act, not the last remaining democratic resort.

Global agreements are big and loose, but they are not enough. The globe is much bigger and more complex than the trading network, but our economic agreements are trashing entire communities and nations, by treating them as sources of cheap labor or materials, then ignoring those without the entry fee in goods. We have trouble conceiving of the complexity of the planet, much less controlling it when combined with the added complexity of human trade and exploitation. Eagles and lichen get pushed to the lowest common denominator. Ecological capital can be ignored. The social capital of communities starts to erode because the most important costs, to communities, members, and the environment, are eliminated or zeroed out in the economic calculations.

Economic globalization is premature and dangerous without a political framework at the international level to moderate the unfair advantages from trade that is not free of profit-focused corporate or political decisions. There need to be international rules and laws to slow the acceleration of trade for profit. They need to be expanded and reestablished before global trade is beneficial to everyone, as it could be.

On the other hand, global politics promises to be a bigger mess if the larger nations are allowed veto powers and special privileges based on some old historical advantage. I fear global politics could destroy any chance for fair representation until the United Nations and the 1000-3000 small nations are empowered. Certainly, it is attractive for big nations to unload the burden of screwing up everything global.

Are Enormous, Simultaneous Global Changes Possible?

The enormity of the military budget suggests that we are willing to dedicate enormous resources for one thing we deem important, although that is not sustainable. However, the military could be assigned to the domestic emergencies at home, for making transitions in energy and agriculture. We need mobilization as fast as happened in WWII. When the auto industry was drafted to make planes and ships. When strategic goods, from sugar to gasoline, were rationed. That emergency restructuring was done in months. Why not try for 6 months for global mobilization, now? Restructuring could be done profitably, to increase global security.

The cost could be quite reasonable. Lester Brown and others have calculated the costs of social goals and of restructuring the earth. The social goals came to $75 billion per year and the restoration goals at $185 billion per year (compared to a world military budget of $1.522 trillion). If we added other social goals, such as equalization and reparations, infrastructure repair, or the deconstruction of some cities and industrial area, it might only double it, to $150 billion. Increasing environmental goals, such as the restoration of farmlands or wilderness might double that estimate to 370 billion. If we added the social costs of transition, including monies for UN to take over police actions, education and health, that might raise social goals to $220 billion. Setting aside more land for hotspots and common areas might reduce income $100 billion per year. The grand total is $690 billion per year.

Compared to $1.522 trillion per year, $690 billion is a half-price bargain. The military budgets are not affordable anyway, which is why social and environmental goals have been so neglected. All that is really required is convincing world leaders that we are in an emergency situation, now, before a major catastrophe claims millions of lives and billions of property. Of course, they would have to sacrifice some money and some comfort, but then they could be heroes. And, that is what we need, uniform sacrifices for the benefit of all.

The expenditures, as well as coordination would have to be handled at the international level, by the United Nations, which would have to monitor progress towards the goals. It would need new offices and agencies to pay out the money allocated for it. The UN could also appoint new Boards to oversea each individual goal. This scale would be larger than anything done before, so it is important to have a hierarchy at the top, but let the smallest groups be loose and inventive. Sometimes the most effective directions come in a bottom-up way.

This scale of change, with actions performed at one time in every community might be possible. Everyone has to understand that it is a planet-wide emergency, as immediate as any hurricane or earthquake; it is just more subtle. There will be social and cultural hurdles. Of course, the rich and powerful now will not want to sacrifice the comfort of the status quo, but they will still benefit more than anyone else. It is in their interest to still be the richest group in a healthy culture and planet. If we can just educate them to their real interests. If. Some cultures will not want to work with other cultures, if they perceive unequal gains. That is why a diverse body of managers is necessary. No matter what the outcome, this has to be tried, for the whole planet not just self-assigned, special groups of humanity.

Ecoforestry Series 1993—

Propose Goals for Global Forest Restoration

Deforestation, with forest decline, is a planetary crisis resulting from numerous cosmological, social, economic, and demographic events. The crisis must be addressed through changes in policies and institutions. Saving the remaining wild forests is urgent, but not sufficient; massive reforestation is required—replanting, regeneration, and the restoration of all forest components, from soil and microorganisms to herbs, trout, and wolves. And, that will require the involvement of communities, in addition to universities, governments and individuals.

Goals, in a sense, are horizons that we travel towards; in another sense they are the tools that shape us as we travel—we are one of the tools, along with natural disturbances, pathogens, and other agents, that shape the forest. Global and national goals are often more important than local goals.

Over 50 percent of the planet was forested at one time—down to about 28 percent now. We could restore at least 20 percent in the next decade, leaving all old growth untouched and only 2 percent for tree plantations. On the other hand, how much forest cover do we need to keep the atmosphere functioning the way we like it? Carbon dioxide CO_2) accumulates in the atmosphere at the rate of 4×10^6 tons annually. Deforestation possibly releases about 2×10^6 tons of carbon per year (only a third as much as from fossil fuel combustion). Reforestation could remove carbon from the atmosphere. We need about 7×10^6 square kilometers of new forest to store 4×10^6 tons of carbon annually (after George Woodwell's estimates). Estimates of the minimum forested area for the planet are more difficult to arrive at. R.A. Houghton suggests that the minimum should be about what remained in 1990: about 5.3×10^9 hectares, or 40 percent of the land area, although not definitely known.

The Worldwatch Institute estimates that to supply current global demands for fuelwood, lumber, and pulpwood, 9 million ha of trees need to be planted each year starting in 1995; for soil and water conservation, another 6 million ha (at an estimated cost of 6 billion dollars); and 110 million ha just to catch up with cutting (these trees would also sequester 700 million tons of carbon, reducing the greenhouse effect).

Global goals include: Reimplement international initiatives to slow deforestation (previous initiatives *accelerated* deforestation); Employ initiatives to reforest areas, such as in Burundi, where only 1% is still virgin; at least 1,296,000 ha of forest should be restored. Plant and maintain forests sufficient to guarantee indefinite support of known and unknown global biogeochemical cycles. Protect fragile ecosystems with global importance.

Reduce threats to forests from acid rain and other nonpoint-source pollutions. For the planet, this means reforesting 1.4 *billion* hectares to restore the 30-40% forest cover removed in the past 3000 years.

Reforestation goals only make sense in the context of other ecological or social/cultural goals, such as reducing demand for wood products, increasing efficiency in use, fitting human populations to biological limits, and educating people about the roles of forests. Personal goals, such as ensuring personal security and fulfillment, living frugally, or questioning industrial practices, are also relevant.

We do not have facts to base our actions on. Nature is a stochastic process, always changing; forests are always changing. There is a profession, however, that acknowledges the operation of chance and makes conclusions in the absence of facts: it is called gambling—few people are successful at it. Restoration and preservation are also gambling. We do not know for sure what effects our actions, even preservation, will have on forests and nature. The proper virtues for gambling with nature are humility and courage, not arrogance and fear. With these virtues, we might develop wisdom.

Calculate the Value of a Forest

How do we even measure the value of a forest—for us, for other species, for itself? Considering economic values alone has become complex. Many economic kinds of value are difficult to quantify. For instance, until people were asked in surveys how much they would pay to have wild forests, existence value (that is, just knowing that a forest exists without having to visit it) did not have a dollar value. Some economists have estimated also numerical amounts for *option values* (retaining options for the future) and *bequest values* (leaving as-is for future generations). For instance, a typical Colorado household was found to be willing to pay annually $4.04 option value, $4.87 existence value, and $5.01 bequest value for 1.2 million acres of wilderness in Colorado.

The kind of value measured most often for forests, other than stumpage, is recreation. The costs of recreation can be identified and added up; although the forest is a "free good," transportation, equipment, or entrance fees are often required to visit it. A typical visit may cost $15.00 a day in 1992 dollars. But this does not reflect the value of the experience. Some economists have proposed a "contingent valuation" by subtracting the actual costs from what the consumer is willing to pay—this surplus value makes the recreational value more objective, because there is a dollar amount.

Another category of economic value to consider is *conservation value*. Natural capital such as forests function to regulate climate, produce topsoil, and cycle elements through the ocean and atmosphere. These environmental services (the wild infrastructure) support the economy without providing direct economic benefits.

Consider, a "what-would-it-cost-to-replace-it" game that analyzes the replacement costs of natural services in terms of human labor and technology. Buckminster Fuller once reported that it would cost just over a million dollars per gallon to manufacture gasoline using chemical processes and electrical power (California Con-Edison 1972 rates). The following list from a 1994 Ecoforestry course is a thought experiment with *very approximate* numbers (from grocery store and internet sources).

Function	Human cost
Pure water	$0.70/liter
Pure air	$0.04/cubic liter
Climate Moderation	$22,000/day
Wind protection	$6000/hectare
Wild genes	$11,000,000/gene
Recreation	$2,000,000/park
Flood control	$24,000/hectare

For a small forested watershed (and airshed), the costs of replacing basic forest 'services' would be billions per year. One thing business can do is put a price on nature, but it should be a real price, reflecting the real cost of replacement; for example, the costs of raising and planting trees, and then monitoring them.

Just last year, at the annual meeting of the American Association for the Advancement of Science, scientists were told that the goods and services provided annually by natural ecosystems are worth many trillions of dollars in conventional economic terms, according to Stanford ecologist Gretchen Daily. These services are the life support functions performed by ecosystems, such as purification of air and water; detoxification and recycling of wastes; generation and maintenance of soil fertility; pollination of crops and other plants; regulation of climate; and mitigation of weather extremes like flood or drought. Ecosystems also provide goods, like timber, whose harvest and trade represent an important part of the human economy.

Ecosystem services operate on such a grand scale and in such intricate and little-explored ways that most could not be replaced by technology. They are priceless, but because they do not have a price, they are not traded in economic markets or considered in calculations. Government and industry urgently need to incorporate these life-support values into their policies and planning. Estimates of the value of nature's goods and services is critical to informing decision-makers.

Yet, it is as impossible to capture the value of a forest as it is to define a human life in strictly economic terms (either as $0.98 or $78,000). Even assigning monetary values to nature, homes and lives is problematic.

Ask anyone who has lost a family member, home, or favorite woods if the money gained was reflective of the real value. Many people do not allow their ancestral homes to be assigned even the highest monetary value, since they have the "right" to occupy them. For example, Thai *muang faai* communities refuse to let sanctions against cutting trees in community forests be interpreted as prices on the trees, which are necessary to supply water to rice fields. Many people in industrial countries refuse to participate in questionnaires that ask them how much money would be acceptable to compensate for the loss of visibility by air pollution or the loss of forests by clearcutting. Refusal to discuss the price of a forest is legitimate communication; it should serve to channel forest policy into larger noneconomic realms.

Even if we cannot put a discrete amount on aesthetic values, common mathematical sense tells us that as the dividend (number of forests) approaches zero, its worth goes to infinity—no one can experience it if it is gone. The infinite value of forests is only a constraint on economic exploitation and reasoning.

All the true costs of any technological process must be internalized even if it means assigning arbitrary dollar values to resources. This does not necessarily mean that an aesthetic resource is worth $1 billion, but that there is a high potential economic and spiritual loss to the system if the resource disappears. Value systems are the driving variables in all economic systems, not just peripheral attachments. And patriarchal nations are devaluing these values and cultural wisdom with their dominant, abstract, quantitative economics. Economic objectives have to include new concepts of value.

We get our values from knowing what is valuable in nature. Values usually encode information having survival or prestige importance. Perhaps the most valuable thing is living time, the experience of life—aesthetics (from the Greek meaning perception). This may be why humans value walking in the woods and acquiring rich sensory experiences from direct contact. But economists rarely mention these values. Light, wind, dirt, plants, birds, all act during a walk, but not with the meaning of crops or sheep, which is for their utility—they just are. People do not live without these things. All values are based ultimately on a healthy ecology. It must be kept healthy.

Figure. Altazor Forest Survey 2000.

Fun with Limited & Imaginary Numbers for Exploitation

The estimated fiscal impact from Initiative Petition 20 in Oregon looks bleak, but these numbers are very limited. Elsewhere and elsewhen one can imagine how other fiscal impacts from other declining industries or one-crop efforts looked equally bleak—for instance, slavery, cotton, beaver pelts, buffalo tongues, or tobacco. Nevertheless, society learned to cope with change and even to prosper.

  I can find no fault with the numbers as they are, but they do seem so isolated on the pages. The total impact is estimated at a negative $74.59 million. What would the numbers look like if we continued clearcutting for 10 years? The income looks good, but the real costs seem extremely high and make the whole sum negative.

Overall timber *income*	+4,000,000,000
Tax revenues	+400,000,000
Legal fees	-95,000,000
Earth First! employment	+1,000,000
Restoration employment	+75,000,000
Costs of roads	-25,000,000
Loss of infrastructure	-25,000,000
Loss of human life	-90,000,000
Loss of jobs to machinery	-13,000,000
Decrease in prices	-15,000,000
Loss of nontimber products	-38,000,000
Loss of recreation revenue	-3,000,000
Loss of amenities	-90,000,000
Loss of community wealth	-1,000,000,000,000
Loss of species/diversity	-1,000,000,000,000
Loss of ecosystem services	-2,200,000,000,000

That is a *minus* $4,195,919,000,000 or over -$4000 billion per year. Perhaps if we used a different system of harvesting, such as selection, or a different kind of forestry, such as restoration or ecological forestry, the numbers would look very different and the sum would be positive:

Overall timber *income*	+2,900,000,000
Nontimber income	+3,000,000,000
Recreation income	+4,000,000,000
Tax revenues	+900,000,000
Legal fees	-5,000,000
Earth First! employment	+500,000
Restoration employment	+75,000,000
Forestry employment	+200,000,000
Costs of roads	-5,000,000
Loss of infrastructure	-2,000,000

Loss of human life	-9,000,000
Loss of species/diversity	-203,000,000
Loss of ecosystem services	-600,000,000

Quite *positive* actually, at $10,251,500,000 or about $10 billion. There might still be some losses, but they would not overwhelm the benefits. These numbers, with a positive sum and not a negative one, would be consistent with the clearcutting ban and indicate the possibility of the indefinitely sustainable use of healthy forests.

Base Forestry on Community Productivity

For one simple experiment, consider the forest as a corporation. After all, politicians and law enforcement officers are impotent to stop the destruction of forest ecosystems. The simplest way to give the forest a voice in its development is to incorporate it following international law. A corporation is just a legal entity with its own rights, privileges, and liabilities. Although a corporation is independent from its founders, it is a human construct and the forest corporation would have to have human representation in the human system (a permanent site forester would probably act in the best interests of her home). The forest corporation would not be really different than most corporations. Like other corporations, its primary purpose would be to maintain its own existence and maximize its wealth. It would optimize its values, which would include tree and fungus values, as well as human ones. This strategy would solve the problem of cutting too much of the forest— the forest itself. The Net Primary Productivity (NPP), would be untouchable capital. The Net Ecosystem Productivity (NEP) is the operating cost for animals that feed on and distribute materials for forest renewal. Dipping into NPP or NEP would reduce the forest itself. Most of the shares would be treasury shares; anything more than par value would go to capital surplus to be distributed as dividends—the dividends would be the net community productivity (NCP).

The NCP is profit. The NCP of forests of varying maturities varies itself from almost 0 percent to over 10 percent. The most NCP one would expect in an old-growth forest would be 0-2 percent. In a young Ponderosa pine forest the percentage may range from 5-10 percent. In early seral stages after some catastrophic change, alder growth may produce 15 percent NCP. Some energy-subsidized pine plantations in Britain exceed 30 percent for a short time. Humans can take part of the profit, the NCP. For example, in a Ponderosa pine forest of 2200 hectares, with an NCP of 2100 kilocalories per square meter per year ($2200 \times 2100 \times 10,000 = 4.62 \times 10^{10}$ Kcal per forest), the NCP is equal to $1. \times 10^7$ kilograms of dry weight, which is equal to 4.1 million cubic meters of wood (weight of Ponderosa pine is .41 kilogram per cubic meter). Market value would vary, but it could top $554 million.

As long as humans limit their take to the NCP, the forest is truly sustainable. If the human managers take part of the NPP, they compete with other animals in exploiting forest resources—competition and exploitation are healthy, remember, so that may not be too bad; but, the forest may not be sustainable indefinitely, unless the human managers replace some of the same functions as the creatures they are competing with. And if they take all of the NPP and most of the GPP, then they interfere with the operation of the ecosystem.

Obviously, we could cut any forest at any rate—we have been doing so. But, the rate at which we cut determines what the forest will look like over time. For instance, if we were to cut all forests at a 1 percent rate, then they would all probably develop old-growth after several hundred years. If we were to cut forests at a rate of 10 percent, some forests would never develop beyond an early stage of maturity. Somehow we need to fit our needs into the production of the forest without interfering with it. We should optimize cutting instead of maximizing it, harvest a percentage of the natural interest instead of the ecological capital. We could encourage diverse forests instead of single-species, even-aged plantations.

The forest is a web-like system that produces many things that are useful to human beings. Limiting the practices to net community productivity or even a percentage of net primary productivity would help save forests, but it would require better planting and restoration of clearcut areas, ecological planning and management of national, corporate, and private forests, improved use of wood products (through reduction and recycling), and a re-evaluation of human needs.

Why is Modern Economics So Clueless & Dangerous?

The myths of the mutant modern economics have tremendous impacts on how forests are treated. They are based on a cosmology of the machine. The old analogy of the economy as a machine leads to dangerous assumptions about forests:

- Everything is a resource (and its corollary, everything has a price [and its corollary, everything that does not have a price is worthless]): Every forest can be cut to provide wood for human needs. The essence of a resource is that its existence acquires meaning only as it is necessary for human needs and wants.
- Resources are unlimited: If forests are unlimited then we can keep cutting them—and even if they are not unlimited, by the tenets of modern economics, being a scarce resource makes them even more valuable.
- The economy has to keep growing to survive: Growth means that the

use of nature will have to keep increasing. If there are not enough plants then we will have to grow fiber in vats. Or, we will have to substitute, for example, steel wall studs for wooden ones.

- Any resource can be extended through a substitute: Therefore forests are not unique or intrinsically valuable, because they can be substituted with tree plantations or genetically engineered industrial wood cell production or steel or plastic.
- Mass production is most efficient: Clearcutting (the incarnation of mass production in forestry) a forest is the most efficient way to extract good wood; high-value native trees are cut without undue expense, while the site is automatically prepared to host a plantation containing only desirable market species, such as Douglas fir. Accounting is also simplified—noncommercial species are considered a waste of potential space.
- Obsolescence is necessary for successful growth: Obsolescent products made from wood are burned, buried in landfills, or stored in concrete hallways, removing them from the cycle of renewal of forests.
- Quality does not matter very much: Good materials, good wood from old-growth trees, are not necessary to make high quality goods. The quality resides in the perception of the consumer, educated by helpful advertising.
- The future is less valuable than the present (discounting): A forest that may have some value now in a few useful species is worthless beyond a certain time (two years, maybe 20). Modern economics has enshrined this one form of selfish behavior and pretends that it is rational and optimal, even if it is inconsistent with the long-term best interests.
- Economists can control the economy: By simplifying the forest and raising one species on a tree plantation, economists think that fertilizers and pesticides can control all foreseeable circumstances—changes in and shifts of the system state can be controlled.

These dangerous assumptions are translated into unhealthy and unsustainable practices that also generate problems, from which the forest economy is suffering. Overgrowth is one problem—too many trees are cut because of demands for wood. The complexity of the system and the number of costs increase; this is only a surprise when total cost accounting is not used. Economic and social instability result from mechanization and the quest for a narrow mechanical efficiency; powerless families are relocated or dislocated as jobs are eliminated by ruthlessly efficient and sophisticated hardware. Forests experience environmental wobble and ecological instability as they are cut; atmospheric and hydrological cycles are simplified and disrupted. Finally, misdirected effort on ill-conceived products, e.g., upscale firewood or throw-away chopsticks, wastes precious wood. The economy cannot adapt to the time scale of the forests or escape its assumptions, so it turns to planning to solve the problems.

Use Gigatrends to Design Forestry

A number of trends in forestry are evident. Some are megatrends; for example: Forest companies have learned to restock some sites, but not to plant well or to nurture the trees; furthermore, rather than moving from industry to information, forestry is moving backwards to providing raw resources; and rather than moving towards multiple options, forestry is backsliding into one option: logs.

The real long trends in forestry—and in forests—occupy the entire human calendar. These gigatrends are larger and more involving than megatrends. There are economic and ecological trends related to forestry, for instance, shortages of timber. Most Old-World civilizations faced shortages of high-quality wood, from the Mesopotamians, Egyptians, Greeks, and Romans to the modern countries of Europe. Pressures on British woodlands in the 1300's forced people to turn to coal as a fuel source (a source regarded as inferior to wood). The timber famine reached Europe in the 1700's. It had existed in China and India over a thousand years before. Countries that exhausted their wood supplies had to invade other countries or find substitute such as coal or water power. Each substitute required more energy to produce. E. Eckholm warns that the serious firewood shortage over most of the earth due to population pressures on the remaining woodlands. There is an overall growth in wood consumption, and an increasing per capita demand for wood.

Ecosystems are simplified and degraded; deforestation, desertification, and exotic take-overs occur on a large scale. Humans simplify ecosystems and keep them at early seral stages to harvest the increased productivity. Much vegetation becomes a social artifact. In Scotland, for instance, forest cover was reduced from 55% of the total area to 5% by primitive stock-keeping and agriculture; the moors decreased by half, but meads increased eight-fold.

Forest cover has been reduced since 1100 B.C., constantly reduced since the 1500s, and drastically reduced since the 1950s. There is little idea of the rates. Overall the world forest cover has been reduced over 35 percent. The actual amount is very uncertain, since inventories are rare or crude, and half the land reported as forest land in many countries is also labeled "unstocked," according to E. Eckholm and the World Forest Inventory. The result is grasslands, scrublands, and wastelands labeled as potential regeneration areas.

The long-term health of forest ecosystems decline as people fight over access to specific resources for short-term economic gain. Natural regeneration is declining, due to interference in the operation of forest ecosystems and the destruction of some of the necessary structure, for example, clearcutting kills mycorrhizal fungi necessary for nutrient uptake.

The intent of describing large-scale patterns is to have human patterns fit with observed patterns in nature; patterns have a form, sometimes repetition, and sometimes regularity, but each of these is caused by some limiting factor, such as water, temperature and soil. Fitting the pattern can lead to both continuity and predictability, and both of these are needed to adapt human activities to natural limits.

Thinking we have conquered nature and are omnipotent, we have quit thinking. Satisfied with our comforts, we do not ask enough of ourselves. With these gigatrends possibly ending in tragedy for humanity, we must ask questions. What kind of forests do we want? Even or uneven-aged? Wild or domestic? Managed or unmanaged? How shall we use those forests? For wood products? To protect watersheds and maintain global biogeochemical processes? As a home for other beings? Recreation? As some kind of balance? These questions lead to new strategies for living with the forests.

Identify Limits for the North Slope of Alaska

This experiment recommends a series of goals and limits in a comprehensive plan for wilderness based on the biohistory of an ecosystem, the cultural values of the people, and knowledge of the limits for sustainable development. This makes the limits explicit and sets equitable goals within the limits. A synthetic framework provides for the health of the ecological system as well as for the health of its human inhabitants. This relates wilderness to resource use by the cultural values of an optimum human population within a home ecosystem, although that system is connected to others by trade for some necessities or luxuries. Cultural goals are keyed to the traditional idea of physical limits.

An optimum amount of wilderness is calculated to preserve the natural cycles indefinitely. If the current area is less than the calculations, the difference is restored and set aside as a reserve. The remaining areas are zoned for appropriate use, including conservation, preservation, and artificial areas (with historical, cultural, and functional importance). Resources available for human use are identified, including raw materials and the usable biological productivity of the areas. Productivity is used to calculate a base line population. Cultural modes—where culture is an ecological activity—set limits on the use of technology and resources. Cultural values drive changes.

Because it will always be subject to human influence and revision, wilderness cannot be considered apart from population growth and human activities. The Inupiat have no word for wilderness; for them the land is peopled with reasoning, speaking beings. At low population levels and high dependence on the resources of the land, it is unlikely that the Inupiat would destabilize the North Slope by their hunting activities. The North Slope Region is already divided into major areas. The National Petroleum Reserve, established in 1923, covers 96,000 km². None of the Petroleum Reserve has been designated as wilderness. Yet.

The recommendation for a North Slope Region reserve is 2 large areas, about 80,000 km² each, coinciding with the two largest caribou herd migration areas, 80,000 for the west Arctic herd and 80,000 for the Porcupine herd; these sizes should be adequate to preserve the associated predator populations as well. They would be buffered by native cultures. The hunting and gathering lifestyle requires huge expanses; the reserve recommendation for native cultures is 100,000 km². The reserve would be managed by benign neglect—despite the fact that many do not list this as a management option. Natural processes, such as fire, wind, or species eruptions, would be allowed to operate freely, even if they altered the functioning of the system. The large areas of the proposed reserves would be laid out on top of the petroleum reserve and the wildlife refuge.

The north-south orientation for the reserves would minimize the dangers from physical and climactic changes. The greenhouse effect could drastically alter the species distributions in reserves, with the loss of many species. Placing the reserves on heterogeneous soil types and topographies increases the chances that the temperature and moisture requirements of species would be met.

Ultimately, North Slope civilization depends on wilderness—for recycling, pest control, genetic diversity, soil-making, and water purification, among other things. North Slope civilization depends on native cultures for vitality, and those cultures depend on wilderness for their context and imagery. Cities and manufacturing can be sited carefully at the edges.

Allow Death in Forests

Death is usually not addressed in scientific literature or research, other than as a limit or a temporary medical shortcoming. In fact, a forest is as much dead as it is alive; there is a rhythm of death and replacement from the cellular to the ecosystem level. Trees in a forest are always dying, either individually or in groups, waves, cohorts, or systems. Forests also may die, if enough of the trees die, as a result of catastrophic change or of too rapid a change. Mortality is a normal part of the life cycle and usually occurs from a combination of factors. Lightning and wind cause tree death and injury. Injury and disease cause many tree deaths. The causes, rates, and patterns of death in tree species are poorly known, according to Jerry Franklin, despite a hundred years of research. That presents a problem: If we do not know the normal rates of mortality in a forest, how will we recognize abnormal ones?

Part of the life cycle of a tree is death. The dead trees keep contributing to the life of the forest, standing for a while (1 to 150 years), then falling and decaying (20-200 years). A certain percentage of death as the normal condition, necessary for the renewal of the forest. The rate of death per year in a mature forest is remarkably constant at about 1-2 percent, even with wind storms, fires, disease outbreaks, and animal damage. In some cases, a larger percentage of the forest is affected. For instance, high elevation balsam fir forests are subject to bands of dieback that progress up the slopes parallel to the contours of slopes. These "fir waves" seem to be triggered by cold winds striking exposed forest margins. A new stand regenerates where the trees have been killed.

Rotting and burning are an integral part of the cycle of life and death in the forest. Tree mortality from diseases rarely removes trees at greater than 1 percent, and occurs on various scales: Gap phases (small scale), forest development (large scale), and landscape patterns (immense scale). Pathogens are one of the determinants of growth and development. Their effect on the long-term health of a forest can only be regarded as positive. Even catastrophic disturbances like hurricanes rarely damage more than 5 percent.

Disturbances that are not part of the history of an ecosystem may cause irreversible changes, because the system has not evolved a defense or response mechanism to a rogue disturbance. A meteor strike would be such a disturbance, especially if the landform was altered by a crater. Human disturbances, in the form of acid rain or clearcutting, are both novel and threatening. If they are small enough or rare enough, however, the ecosystem may rebound.

Very large scale disturbances, such as volcanic eruptions or meteor impacts, can destroy entire ecosystems or disrupt global biogeochemical cycles. However, such very large scale disturbances are rare, and the ecosystems often have thousands or millions of years to become reestablished, although changed. Natural disturbances, from the death of individual trees to large fires or windstorms, are responsible for critical composition, structures, and ecosystem functioning necessary to maintain fully functioning forests. For example, the death of an individual tree sets off a process of change, beginning with a standing snag that provides habitat for cavity-nesting birds and ends with a fully decayed fallen tree that serves as a water storage and filtration system.

Health is the coherence of the pattern of living in other words. If the pattern is disrupted, the local entity dies. Many local patterns flow together through time interdependently, sharing materials. The death of one pattern sometimes leads to the death of other patterns. The body of a forest or any entity is a dynamic pattern supported by dynamic processes that include other entities. Perhaps this is the best working definition of health is harmony, and that requires death. The universe comes with the whole package and we risk serious error by choosing only the tendencies we want.

Waldgedankenexperiment—Forest Thought Experiments

Humanity is engaged in a great experiment with the planet's forests; unfortunately the experiment is not only bad science (no control planet), it is ill-considered (insufficient knowledge or time to repeat). We are replacing large old complex forests with young simple fragments, in which fires are suppressed, large predators are removed, large herbivore populations are encouraged, exotic species are introduced, soil is compacted, and excessive biomass is removed. Our actions are experiments, whether we want them to be or not. Ignorance, denial, virtue, or cupidity cannot unmake this experimental course, which may be global and irreversible. There must be a way to refine the experiments, to minimize our impacts and to anticipate the outcome of them. Perhaps we could imagine experiments to help us understand what is happening in and to forests. The thought experiments presented below for the USA are incomplete, but suggestive of what we could do or should not do.

1. What if the US Forest Service *reopened all sales* with new rules. Interested parties would make proposals rather than bids. USFS would sell timber rights for 100 years. The buyer would pay the USFS fair market value for the timber. The size of the sales would be determined by habitat type, so there would be an equal number of small sales. The buyer could then log within ecological limits or preserve the area, depending on whether the buyer might be Georgia Pacific or Greenpeace. The buyer would be responsible for building roads, if any, to USFS specifications at buyer cost. The buyer would only pay taxes on detrimental environmental impacts, such as erosion, loss of fish streams, or aesthetic loss—no taxes would be paid on production or profits. Would this work?

2. Simplify forests by removing species. What would happen if we *removed species one at a time,* tracing all its connections, instead of removing the entire superstructure? Would we identify keystone species (or keystone mutualists or key linkages or taxa or processes) as things collapsed? Possible keystones include: Fig trees in Amazonian rainforests, whitebark pine in boreal forests, beavers in North America, termites and elephants on the African savanna. Would each species be found to have a unique function in the community? How much redundancy would there be? What if we removed different amounts of biomass in similar forests on each rotation—at what point would the forests fail to return?

3. *Model the forest* by totaling all the behaviors of individual beings—some computer modeling has started to do this with just trees, and although that is one way of getting some information, a forest is not just trees or even a collection of all individual beings. If we could add up all the activities of individuals (with enough labor and time), we would have a lot of specific information about that forest. Very little could be applied to other forests,

in other places, with other components, unless we were able to make general conclusions. Would the general conclusions be applicable to other forests?

4. *Replace forest functions with mechanical devices.* The machine metaphor dominates modern forestry, resulting in tree plantations. On some tree plantations, many of the functions of a wild forest have to be ingeniously and expensively duplicated. For instance, shade cards are used to protect young shade-intolerant species; plastic sheaths are used to protect bark from predators; fertilizer is used on young trees; and some trees are doped with mycorrhizal fungi. Extending this trend, modern forestry eventually may try to create an artificial forest with just one living organism, Douglas-fir trees. For example, we could replace the functions of nurse logs with gigantic nylon sponges. What artifacts or tools would we invent to replace the functions of woodpeckers, bats, insects, fungi, shrubs, or snags in a mature forest? Microbots or avibots?

5. *Costing out replacement of services.* Suppose that we had to replace every function the forest provides with a human service. Could we afford it? As calculated elsewhere, for an area of 50,000 people, the costs of replacing basic forest services could be billions of dollars per year. Many of these functions, like clean water, seem affordable to replace, but many such as climate moderation or soil creation are not affordable in any practical way.

6. What if we got tired of all that work and decided to *save half the planet in wilderness?* Imagine 6 billion square kilometers of forests set aside, with another 9 billion given over to the control of archaic cultures who reside in them. What changes would we really have to make? Substitution? We are good at that. Higher density living in cities? That trend has been evident for over a century; why not just plan it properly?

7. *Imagining a planet without trees.* Imagine that forestry and conservation both have failed, and the earth is a planet without forests and without trees in general—except for a few artifacts kept in arboretums. For the first time in over a billion years, the planet is not sheltered, the climate is not moderated, and other plants, animals, fungi, and bacteria are not protected. Humans have made shelter for themselves, but we do not want to share it with mosquitoes or grizzlies. Have the ice caps melted? What is the shape of the global system? Are all human crops grown inside? Is the reduction of biodiversity causing unalterable changes? Can the forest ever be replanted? What kind of forests could live without fungi, bats and centipedes? Imagine what changes would occur in human psychology. Would planting trees be required by law in our treeless world?

8. Forests have been changing for millions of years. *A macrochronoscope in low orbit* over the planet for the last several billion years would show forests moving like shadows over the landscape, with climate changes, glaciers, and shifts in moisture. The forests are connected to the land, air and water; everything is constantly changing in a gigantic intricate web. The immense patterns are easy to detect with this imaginary device. On the

ground, changes seem chaotic—forests are chaotic systems. Order and chaos seem contradictory because of our linear thinking. Linear thinking can be illustrated by industrial forestry: if something works well, such as cutting old growth and planting Douglas-fir, then more is better. Nature, however, is nonlinear. Maybe we should modify our logic and our behavior.

We could set up large-scale, long-term experiments, but they are expensive and relatively few. Most experiments are short-range, small-scale, isolated, and detail dense. We cannot experiment at all in a traditional sense, where we hold most variables fixed, while changing one or two in experimental runs. Forests operate over very long time spans; furthermore, their historical nature means that they cannot be restarted.

The best response to a complex question is a thought experiment. Thought experiments can help us avoid being overwhelmed by details, and help formulate goals and interpret information appropriate to scale. Thought experiments can give us clues about what can happen—"And then what?" as Garrett Hardin always asks—and what is the likelihood of it happening. Unlike medical doctors or scientists, we cannot either wait or directly experiment in a realistic time frame or scale.

Thought experiments are vital to understanding the complexity of forests. In practice, erring on the side of preservation—the prudent and conservative course—means minimizing the influence of human activities on the land. It means experimenting cautiously with new approaches to forestry and being properly skeptical about claims for sustainability. It means drastically reducing our demand for wood products, through conservation, reuse, recycling, and human population control, so that the greatest possible amount of natural forest can be left wild and degraded forest lands have time to be restored to health.

Figure. Prescribed burn at MGC. Note Firetruck.

Practice Wild Thinking

Are there large-scale forms of thought that could address large, long-lived entities, such as forests, ocean or the planet? Is Philosophy one such form? Religion? Ecology? What is wild? That which is not domesticated? Uncontrolled or unmanaged? Not cultivated? Untamed? Savage? Wasteful? In a state of nature? Lawless? Wild as a word is ambiguous and reflexive.

What is thinking? To revolve ideas in the mind? To design or to imagine? What is wild thinking? Is ecological thinking intrinsically wild? Is ecological knowledge? Will Wright suggests that knowledge becomes wild when it is critically reflexive and committed to critical access rather than to a version of absolute reality. Thus such wild knowledge cannot be "domesticated" by one particular social institution. It is accessible by individuals.

Wild thinking might ask, what is the proper language for humans? What is the proper diet for all humans? The proper mode of expression? These questions may be too presumptuous to even be asked. We could start by asking if there are different forms of wild thinking.

Ecological thinking is wild because it has a nonhuman component. Wild ecological thinking combines technique with moral and ecological concerns. The fundamental emotion of wild thinking is astonishment, literally being struck by lightning.

By contrast, scientific or religious thinking is domesticated or tamed because it is limited by a human definition of a true reality, or a set of rules for observing a true reality. Science is defined in opposition to religion, with a commitment to neutral observation, rather than a moral commitment to tradition, but their limits make them tame. Tame ideas are remarkably persistent—e.g., "more is better." Many tame ideas, such as "humanity can control nature," should have an expiration date. George Orwell referred to these obsolete ideas as "wrong-think." Another phrase, "double-think," is used traditionally to keep some ideas tame. Gordon R. Taylor suggests that "Non-think," the failure of good ideas to be recognized or used, is equally obstructive—thus the idea "protect the ecological basis of life," is never seriously considered.

Wild ideas can shake up tame ideas. Wild thinking is appropriate for "system breaks," the social discontinuities identified by K. Boulding. We have started to identify the forces setting up the next big break for civilization, but we have not defined the forming patterns very well. We need to be rethinking—a form of wild thinking—the basic assumptions of the spheres of civilization, from our economic and political to industrial, religious and scientific.

Perhaps, though, wild thinking has drawbacks. Too much anarchy is dreaming; too much feral thinking is noncultural and dangerous. Too much wild thinking is unrelated to the important mode of learning by doing.

Ecology is a study of relations; there are too few relations now between honor and community or between work and reward. Ecological thinking could unearth those weak connections, for people to strengthen.

Is wildness just the nonhuman part of the spectrum? Does it overlap in humans? Is it just difference or craziness? We love and celebrate the wild, but we also fear and suppress the wild. The wild is a quality of being just beyond rules or outside of walls. Paul Shepard reminds us that wildness occurs in many places, in any species whose sexual assortment and genealogy are not controlled by human beings. Darwin reminds us that humanity is wild, also, with the possibility of a wild, savage mind. Can such a mind be only wild? Not necessarily—we can domesticate some of our wild ideas. But, maybe we should rewild some of our domestic ideas. Is it meaningful to talk of a wild culture, one that intermeshes with the wild of nature? Are archaic cultures wild? Would a wild culture be a model of a civilization without walls, without resources, maxima, or weeds?

Do Good in Forests

Philosophers have puzzled over the term 'good' for centuries, constructing partial theories and contradictory systems. According to Plato, technical knowledge is not of ultimate importance for human beings because it knows nothing of good itself. The word 'good' has an interesting and long history. The current version is derived from the old English, 'god,' meaning suitable or fitting. Perhaps, as a working definition, we can just use harmony. In Chinese medical tradition, the highest good is harmony, especially social harmony or good relations. A good person creates and maintains harmony. Harmony is related to wholeness—the word 'whole' comes from the Indo-European root which is the root for 'health' and 'holy.'

Still, the use of the word good with forestry is problematic. Good standards or codes mean different things to different people. Furthermore, we cannot know, or even think of anything, according to Robert Zajonc, without some involvement of emotion, that is, at least a vague feeling of good or bad. There are questions of what one ought to do.

Some instances of different kinds of forestry practices can be called good. Thus, industrial foresters as well as small land owners and swiddeners can practice good forestry. Perhaps good forestry is nonsense without the recognition of intrinsic value, but good forestry can always be practiced, at least on a small scale, independently of notions of intrinsic value.

Any example of forestry can be good, if it follows its methods, regardless if it is totally external, human-centered, or simply pragmatic, *if it does the right thing* for the forest—the right thing being an absence of interference with the dynamic systems that shape and maintain forest processes.

Good forestry can happen for the most contradictory or trivial reasons: Self-limitation, true love for the forest, accident, intention, techniques, or shame—forest management sometimes gets altered for the better after public interest and scrutiny.

Good or bad practices change with advancing knowledge. Foresters used to recommend cleaning the trees out of a stream and removing flammable brush and woody debris from the forest floor. Now, they advise us to drop trees into the stream and leave all the woody debris and brush on the forest floor. Which action is good? Or bad? Unfortunately, the outcome of our exploitation or interference may not be evident for hundreds of years. Perhaps we should aim for harmony, for the health of forest and human communities.

To be a good forester, you almost have to be part of a good culture, and that culture must almost certainly have a good image of the world. Otherwise, the culture may destroy whatever good work you do. Or you, as a good forester, cannot just care for the forest. You must also participate in society, and if it is blinded by greed and bad ideas, you must try to influence it to the good. Your being good may contribute to a social good.

We could stress harmony, for the health of the forest, not because the action is good. In doing, we choose between good and bad actions; judgment makes us susceptible to error. Ernst Becker concludes that we find things bad when they are built only for utility. Good forestry can do nothing more. Perhaps, as John Fowles suggests, all our judgments of good and bad are meaningless in the long run. All actions, good or bad, interweave so extensively as time passes that their individual goodness or badness disappears. Judgments evaporate and landscapes remain. Even so, we must act, guided by notions of goodness to set a direction.

More than creating good images and good goals, we should be concerned with resacralizing landscapes, with restoring them to their extents and grandeurs, by regrounding science in ethics, that is, ways of living together, and by changing our attitudes from utilization and flat efficiency towards awe and appreciation. Our detachment from trees and other beings has to end. Our participation in the life of the forest must begin.

*Figure. Restoration at Mountain
    Grove Forest, Oregon
    (Credit: MGC).*

Reforest Ireland

This experiment proposes to redesign Ireland with a comprehensive regional ecological design process based on the biohistory of its ecosystems, the cultural values of the people, and knowledge of the limits for sustainable development. Creating this regional design requires common sense, as well as knowledge of ecological design, critical thought experiments and eutopian strategies. This approach makes the ecological and cultural limits explicit and sets equitable goals within the limits. As a synthetic framework, it provides for the health of the ecological system, as well as for the health of its human inhabitants. This design considers the whole system and designs communities for an optimal fit within the limits of the system. It is a conscious adaptation of the benefits of technology to the traditional idea of limits.

The pattern of forests in Ireland had been established before 5,000 years ago. The climatic optimum of the Holocene promoted mires in the lowlands and at high altitudes above the treeline, which may have been dominated by Scots Pine, with some sub-alpine juniper or birch scrub. The pine forests were an extension of the boreal coniferous forests of Europe. Possibly 65% of Ireland had natural forest cover (surprisingly less than Europe at 73 percent, but well above the planet average), before large-scale disturbance by human society began. Figures are based on a map of estimated forest cover developed by the World Conservation Monitoring Centre (WCMC). About half of this may have had a relatively open canopy (from 10-20% coverage, a tree height of 5 meters, and with a continuous grass layer underneath) and most of the rest was partly closed (25-50%). There were some scrublands and grasslands. Because of the climate, there were essentially no drylands, deserts or areas of sparse vegetation, or extents covered by ice.

Ireland has been subdivided into a series of physical regions largely formed by the effects of glaciation: The Antrim plateau, the Drumlin belt, the western Caledonian province, the Mourne uplands, the central plain, the southern hill and vale province, the south-eastern Caledonian province, and the Munster ridge and valley province. In the North-Western Caledonian Province, the Ice Age stripped the whole region of soil, and today even the valleys have a lack of soil. The region is thus agriculturally poor, with settlement being limited to the coasts and the upland areas generally being left to small lakes and bogs. Agriculture is possible in these sheltered Glens, but the top of the Antrim Plateau remains largely uninhabited. The drumlins are poor agriculturally, and the hollows between them tend to become water-logged (although the lower rainfall in the east means that this is more of a problem in the west). In the Western Caledonian Province almost all the soil was scraped away in the Ice Age, and the region now consists of barren rock, bog and small lakes. There is almost no human settlement in the interior. The Central Plain is a large, low-lying region dominated by the Shannon basin underlaid by limestone rocks and covered in glacial drift.

The small remaining native woodlands or old-growth forests represent small 'hotspots' of biodiversity, with small habitats for a diverse range of fauna and flora. The native woodlands have contributed disproportionately to Irish history and culture, in numerous references to native trees, place names and song. For instance, the ash tree is synonymous with the national sport of hurling and continues to be part of sports culture. Recently, the People's Millennium Forests Project recognized how important forests are to Irish heritage.

Although virtually all of the original Irish forests have been destroyed, remnants of woods from the Middle Ages still exist. Some ancient woodlands regenerated from the practice of coppicing (allowing the trees to sprout back from the trunk). Population pressure and the constant creation of farmland were responsible for most of the loss, although natural bog expansion continued—the clearing of forests may have accelerated the process based on evidence of stumps in normal soil.

In the Mesolithic, elm dominated many forests, but by the Neolithic the remaining woods were composed of oak and hazel, on land too infertile or too steep to be useful for agriculture. Even these small remnants were replaced by conifer plantations in the 1800s, as part of the modernization of forestry on the German model. The Forests Project in Ireland could tie together different methods and different lines of evidence, from written records and pollen analysis to archaeology and ecological experimentation. The challenge still remains to understand and account for the changes that require longer than a human lifetime.

Ireland has 70,286 square kilometers of land. Urban land is less than half of 1% of the area (and about half of the area in natural forests). Just over 2% of forests are protected. Currently, there is seven times the area of plantations as natural forest. Most of the plantations are pine. Currently, only 1% of the land area is covered by native forests. Over 90% is covered by cropland and cropland mosaic, which includes mixed forests and some coniferous plantations. Currently, no forests or plantations are certified for sustainable practices by the Forest Stewardship Council.

A country's future wealth might be gauged by its tree cover, according the Richard St. Barbe Baker. Ireland is one of the least wooded lands in Europe. Will its wealth decline in 200 years? Baker calculates that the minimum forest cover for safety is a third of a country's land area, more than the current cover in Ireland. Furthermore, he states that about 22% of a farm in shelterbelts could double the yield of the farm. Baker divides agricultural land into seven grades, and forest into seven grades; the last three grades of fields overlap with first three of forest; the lower grades of forest land could be used for agriculture, especially natural farming, permaculture, treenut culture, or organic farming.

Ireland now has a Native Woodland Scheme whose primary objective is to promote both the conservation and enhancement of forest biodiversity in forests. The projects intend to reinstate the native woodland type, characteristics and attributes deemed most suitable for their particular site conditions. Timber production is encouraged, where appropriate, for quality hardwood and high-quality timber. Education and training are important.

This experiment asks what did Irish forests look like? Why did coppicing decline early in Ireland? Why does Ireland have so few ancient trees? The experiment recommends *replanting and restoring over 30% of the land area in forests,* by concentrating on lands with poor soils, on hills and along corridors that would connect the forests across the island in one linked network, which would allow plant and animal movements to resume around the edges of agricultural fields.

The Forest Service is the forest authority in Ireland and is responsible for overall forest policy and production of forest statistics. The Forest Industry and Planning System records the location and broad 'crop' types of all forests. There is a proposed Forest Service inventory to assess the status and trends of Ireland's native woodlands to encourage the expansion of native woodlands and biodiversity through the application of appropriate silviculture. Continuous cover silviculture, natural regeneration and other approaches that enhance biodiversity are specifically encouraged, as are operations that minimize site and habitat disturbance. A research project would investigate biodiversity in plantation forestry.

Management experience is intended to be combined with scientific, indigenous and local information for the expressed purpose of improving practices. This would include a network of Monitoring Sites where different management approaches in various woodland types could be assessed, as well as professional conferences, training courses, workshops, and manuals and guidelines. Management has to address threats to forests. Invasive exotic species, such as rhododendron and laurel, are major threats to native forests. They must be controlled or eliminated so that they do not prevent natural regeneration.

Figure. Forest Restoration Plan for Ireland.

New Forests
Arable, Dairying, Mixed

Forestry is Poetic Activity

It has been said that vision without a task is merely dreaming. But dreaming is a necessary first step before thinking, planning or making. Ecological forestry is literally a remaking of the landscape (the Greek word for making is *poeisis*). We are making the forests of the future with our actions today. We need to use our aesthetic senses to do that. We need to use a poetic process.

Native Americans used metaphors in teachings to show relations and change people. Science itself makes extended use of the metaphorical process to construct its models. For example: 'A tree is a machine' according to David Smith and others or 'The brain is a computer' according to Michael Arbib. This kind of metaphor is used to create our images of nature as resource. The use of the word ecology by Ernst Haeckel implied that the natural world was a place to live, a house, rather than a machine to control. Making the earth into a house is fundamentally a poetic activity, according to Gaston Bachelard. Poetry also is a way of understanding the universe through metaphor, a literary device that transfers the characteristics of one term to another.

Poetry is communicative of the quality of things. Like science, it discriminates the unsuspected in the commonplace. It is not different from science, but more diffuse; not better than science, but more comprehensive. It accepts ontological parity, the equality of beings; aspects of the world are not negated or reduced by one another. As metaphorical knowledge, which may be prerational or metarational, poetry can avail itself still of scientific references. Poetry can measure a whole qualitatively and mimetically, a germ or the cosmos with its imagery. Poetry is a tool for comprehending partially what cannot be known totally. A poetic language could include a view of the interrelatedness of all existence in a sublime ecology.

Poetry does this through metaphor, which can be understood as the connection of a focus and a frame. Focus and frame can be understood together metaphorically. A metaphor can be understood as consisting of two parts, according to Max Black: A focus and the frame. The focus (or figure) designates the figurative term signified through the process, and the frame (or ground) refers to the subject or context. Using this distinction, it can be seen that most of the fuss in forestry has occurred at the focus level. Foresters have so long focused on trees that they forget that the forest is a frame that holds many foci (or points of view).

Consider the importance of patterns in forestry. A pattern can be defined as "process applied to components" where the process is actually prior to the components. The elements of a forest are related psychologically, by foresters, as focus or frame, as contrast or uniformity, as dominant or recessive, or in a number of other pairs. For instance, forests can be considered by scientists as either matter systems or energy systems, but the focus on either frame permits subtle differences and limitations in interpretation. Some ecologists describe organisms as being configured by energy through time, but organisms are material patterns in space as well.

We have neglected slow patterns in forestry, those that move across landscapes over thousands or millions of years. We have neglected the importance of relationships. David Perry suggests the metaphor of sailing for us to consider that foresters are like sailors. This is an excellent metaphor because it reminds us that we cannot control all the forces of nature, only understand them and move with them to get where we want to go. Sail on, and enjoy the ride!

Remake the Self-making: Ecological Design

Nature is self-making and self-designing, but we humans now influence every natural system, taking what we need from some ecosystems, enhancing a few, misusing others, and interfering with the rest. We need ecological designs to restore the balance between human needs and natural processes. Ecological designs focus on whole communities that work in the same self-sustaining and self-limiting ways as nature. By consciously creating meaningful order, we can develop ways of producing widespread community wealth while positioning the community for a long, sustainable future in a healthy environment.

There is no guarantee that nature can provide humans with everything they want. Recognizing the lack of guarantee simply admits that nature is wild and we must come to terms with nonhuman beings and processes. It is not enough to arrange trees in rows to maximize future harvests; it is not enough to preserve small areas of old-growth without natural disturbances. We must pay attention to the processes that make up the habitat, for example, the role of herbivores on trimming vegetation and diversifying it by predation. The design of the forest and its management must ensure that the processes operate to maintain a dynamic state. Furthermore, the context must be conserved. The forest, however, cannot be considered outside of the context of the entire landscape, including human images and institutions.

As design process includes planning of ecosystems, this mean a fourth level of design, the communities in a landscape (the first three are components, products and systems). Many more problems occur at level four (and the global would be a fifth).

All levels of design need to be addressed, from the conceptual to the political, and are involved in all stages of the process. This involves new challenges for ecological design to: Relate a project to its total context; be concerned as much with cultural survival, justice, and wilderness preservation as with efficiency and aesthetics. Consider the whole perspective; the proper vision is of the whole community in which we dwell. We need to apply ecological concepts, such as networks and carrying capacity. Designs need to be anticipatory, flexible, pluralistic, polyvalent, and polytechnic, rising from the genius of place. And, we need to participate in place, care for all inhabitants, and assume responsibility for the designs.

Ecological design attempts to restore balance, and integrate human activity into the larger community. A moderate number of human impacts can be absorbed by the system—too many destroy the systems capacity for self-maintenance. The design should be open to evolution and to human technological and social development. The design should be based on a model of ecosystem functions, considering diversity, complexity, and the maintenance of natural process—natural meaning a self-sustaining system composed of elements now lost through human interference. This also means restoring wild systems.

An ecological design involves designers and people in reshaping and recreating a self-sustaining community. Ecological design is the design of communities. Everyone can participate. We design places as organic wholes to promote the well-being of individuals and the common good. The immediate goals of design are to reverse degradation and reclaim places for communities, but also to work to increase public awareness of the interdependence of communities, to create environmental quality, and to transform public values by generating new metaphors for living. The goal of ecological design is not to restore, but to revitalize and reinhabit ecological communities. We do not want to live in the dead bones of a mechanistic failure. We want to live in a healthy environment with aesthetic appeal.

Figure. Ecological Design DVD cover (Credit: Sundance).

Rewild the Ocean

The ocean is no longer a temple, but the ruin of a temple, beautiful in its reminder of its former glory, but not functional at renewing itself. We can bring back some of the functions and some of the colors, but it will involve sacrifice. We will have to change our diets. We will have to restrain our appetites. We may go hungry sometimes and live on less wholesome and reduced varieties of food.

Because, in order to rewild the ocean, we will have to stop 90% of our fishing. That means no factory ships. It means no ocean aquaculture or mega-engineering of ocean fisheries. It will be hard on cultures that depend on the ocean, but if we don't do this now, the oceans themselves will be harder on those cultures; there will be nothing left to catch!

We have almost ruined the ocean, as we used our finder technology and mining techniques to strip entire fish and bottom communities. At one time, Grey whales swam in Vancouver Harbor; within 40 years they had all been killed. Eighty species of fish are absent from the New York area. The first species to disappear from the Wadden Sea were whales. Grey whales in the Atlantic were wiped out. Then northern right whale. When Antarctica's great whales were nearly wiped out by the 1960s, the krill population plunged instead of swelled. Whales had created a good environment for the krill. In the North Atlantic, biomass has declined by 97 percent. As stocks collapse, fishing collapses.

Paul Watson says we are eating the ocean alive, although fish harvests are not growing anymore. We are polluting it with toxic chemicals and wastes, that also cause acidification, which is hard on shelled animals and plankton. In the 'Economics of Extinction,' it pays to kill everything to justify the technology investment. Furthermore, fish like Bluefin Tuna become more valuable as they become rare. The last fish will likely fetch a million dollars. The concern is profit, not a living sustainable resource.

The damage is so immoral that denial of our slaughters is useful for us to retain our sanity. We are innocent of recognizing the truth. The sense of loss from destructions and extinctions would be hard to bear. But, the knowledge seeps into consciousness slowly. And, the only way to assuage it, is to allow the ocean to rebuild by removing our catastrophic interference. Unlike the land, we do not need to live in the ocean and we do not need build new ecosystems. All we have to do is let it repair itself. We could, and should, restore most of the reefs and shorelines. This is only intelligent self-interest.

No one has developed a model of the historical abundance of the entire world ocean. No way to guess how many fish could be taken below limits of decline. Perhaps in the future, we could inventory the oceans and determine how to exploit a percentage of a population without destroying it or its habitat. In this case we have to conserve the entire ocean ecosystem by withdrawing.

Moreover, rivers play an important part in regional and global biogeo-chemical cycles, by transporting materials to the ocean. Ninety percent of the landmass drained by rivers. Since river systems discharge nutrients, we would have to restore the rivers, many of which barely reach the ocean. This means increasing the water flow to previous levels, by allocating use and if necessary rationing water use.

In ten or a hundred or more years, depending on the ages of some fish and aquatic mammals, we might start taking fish carefully, noting which populations are stimulated or depressed by the percentage of harvest. The catch over the entire ocean may be less than half of the current mass vacu-uming. We will have to reduce our waste and value the fish that are eaten. Perhaps, we have to deny fish to domestic animals as a food supplement.

Define Ecosystem Medicine

Medicine has gradually extended its attention to wider domains, from the personal health of individuals to the health of human communities and the health of those communities in an ecological matrix. It will be concerned eventually with the health of the environment.

The next step is to integrate ecosystem medicine with environmental medicine and ecosystem science. This means starting by creating a body of cases, then, evaluating the responses to the cases, and finally using that body of experience as a predictive for future treatments. A short discussion of ecosystems is necessary first.

Traditional medicine has developed traditional medical specialties. There are rough parallels with ecosystem care. For instance, ecosystem medicine might include: Examination by infrared or other parts of the spectrum (the equivalent to radiology); the study of boundaries, or ecotones (similar to dermatology); immunology (reaction to exotic species or foreign substances); pathology (chemical, virology); internal functioning (metabo-lism, element cycles, infection); surgery (removal of exotics from invaded systems); rehabilitation (restoration of structure or function); and preventive efforts (ecosystem health).

Ecosystem medicine covers the spectrum from human organs to the complete ecosystems of the planet and maybe the planet itself. Everything could be considered nested ecosystems, from a body to the planet. Ecosys-tem medicine is not exactly humanitarian or ecocentric. Although it ad-dresses ethical concerns and the focus is on ecosystems, it is more ecoperiph-eral; it approaches its subject sideways, cautiously and slightly out of focus.

Ecosystem medicine asks a much larger question that is integrative and contextually sensitive: Is the physiology of the base system healthy enough for self-renewal? Focus is shifted from the symptoms of a disease to the entire functioning process of existence within ecological limits. Health is embedded in the context of the system, with all of its constraints, limits, and opportunities for development.

Medicine can be destructive to individuals, land use can be destructive to landscapes, and ecology can be destructive to animals and ecosystems. The concept of ecosystem health, by bringing the pieces together, could avoid much of the destructive behavior of semi-autonomous departments that are not questioned or contained as part of a larger picture.

There are limitations for our treatment of ecosystems. We do not have a list of diseases for ecosystems. It is difficult to see the effects of ecosystem change or disease from the air or from a distance. Ecosystem illness is similar to chronic illness in humans. The appearance of symptoms indicate that the disease began long before the symptoms became apparent. So, the exact date of onset cannot be pinpointed. If so, then prevention becomes more important. And it is related to genetics, lifestyles (of forest entities, for instance), stress, and environmental effects. Furthermore, prevention is not a short-term, one step solution.

Some patterns of resource use need to be understood on the landscape level (and now on the global level). Others need to be examined over a year, decade or life of the system (especially in the Amazonian forest). There has to be a landscape level mechanism (and also a global one now) for regulating common resources. Local diverse patterns are good, but they must be coordinated at the landscape or higher level. It is the normal range of variability that has to be defined. Models are needed for a wide range of habitats.

There must be an institutional framework, with legal and social mechanisms, for monitoring and managing common resources. This is necessary for the health of the system, as well as the long-term health of humans. That means all interests have to be represented, even global corporations trying to profit from disease or rareness (from overuse).

Public health specialists and population biologists look at disease in populations, whereas, doctors and veterinarians consider disease in individuals. The latter intervene while the former observe. The ecosystem doctor has to intervene in ecosystems. Ecosystem doctors care for the basic systems.

Surveys need to be made of every kind of ecosystem. We need to have an inventory of kinds of systems and kinds of changes. How is conserving terrestrial animals part of conserving ecosystem health? We do not know whether animal declines were caused by disease or some other factor (competition or predation). We need to find that out.

Monitoring is the key to understanding changes. Disease needs to be monitored as an important indicator of integrity. Other indicators are surveys of key species, habitat mapping and human impacts monitoring. Complex interactions have to be monitored, using a range of indicators at levels from behavioral to ecological. There may be limitations of the bioindicators of ecosystem health. Perhaps we need to find common and endangered indigenous species and monitor them, hoping that would reflect the health of the system.

Many systems that have been overused to collapse need to be restored. Especially in forests, many small changes have a cumulative effect. Climate change can lead to loss of forest biomass. Change of rainfall patterns lead to water stress in adapted species. Restoration is a necessary part of medical treatment.

Apply Ecosystem Medicine to Restore Health

Less than 30 years ago, the environment was of little concern to most people. Now it is the primary issue for most people. Calvin Coolidge once said that the "business of America of business." The biggest single business in most of the world, now, is the environment. All farming, most pharmacology, most tourism, and many other "industries" have their basis in the health and beauty of the environment. The environment contributes to the largest share of most gross national products directly or indirectly.

Historically, we have used ecosystems without regard to their continuity or to their health. Partial knowledge and technology has allowed us to exploit our environment beyond what is desirable for us or for other species. While moderate exploitation is necessary to live, too much exploitation is unwise. A wise use of resources would not make the world less habitable. We are part of the system and must protect its health as a whole.

A new category of medical professional is needed: People who address the health of ecosystem themselves. Human physicians may need to be able to identify critical environmental conditions that affect human health, but others are needed to identify the health of those systems themselves. Human physicians need to know the basic principles of diseases related to environmental change or chemical exposure; others need to know the principles of ecosystem health and how that is related to human health.

Conventional medicine has great strengths, as do alternative medicine or traditional cultural medicines. Ecosystem medicine must develop such strengths. Ecosystem medicine incorporates all other kinds of medicine as special cases. It can use the best and most appropriate of any procedures. No single approach works best for every instance.

Using a model of ecosystem medicine offers a comprehensive framework for investigating the problems of health, from individuals to ecosystems, especially the interactions between individuals, social groups, place, and environmental change.

Ecosystem medicine is a medical discipline, aimed at restoring ecosystems to health. As with any medicine, the patient actually does most of the work to become healthy, although the doctor gets the credit and the payment. This leads to respect for the practitioners, but also to more responsibility and more rules. The first rule, which we might take to be basic, is identical to the first vow of the Hippocratic oath, "Do no harm." Noninterference is a basic ecological principle—do not interfere with the stability of the ecosystem—the health and stability of the ecosystem come first.

Ecosystem medicine bases itself in a community context and limits the use of the ecosystem to that which the ecosystem can afford to provide and remain healthy over indefinite time. Ecosystem Medicine undertakes the responsibility to preserve the healthy functioning of the ecosystems under its domain. It also has a responsibility for the ecological production of goods from an ecosystem.

This is the local application of ideas for specific ecosystems, which may have a direct link to human health. The systems still have to be self-sustaining and self-renewing, Of course, the planet can also be linked directly to human health. In traditional medicine the organs are more integrated than in ecosystems, which are looser and more complex. James Lovelock has emphasized the health of the planet as a single system.

The health of ecosystems and human institutions should be measured with a holistic index. We have not developed qualitative indicators of ecological health or quantitative measures of social health, much less an ecocentric view that would value preserves of nature for themselves. To address the health of ecosystems, ecosystem medicine would be a temporary medicine, not a constant intervention or even a continuous diet. We have already tried to gain complete control over ecosystems through scientific methods and technological applications. We have not been able to control them successfully. We regard modern medicine as a foolproof system that tried to eliminate weakness, disease, and mistakes. It has not. Ecosystem medicine must limit its goals.

One goal is the pursuit of the health of ecosystems and their inhabitants. That goal is good health. Individual health is in the context of community health, which is in the context of ecosystem health.

Another goal is security. Symptoms of insecurity are poor human health, migrations, and conflicts (territorial or religious). Environmental security is more than the abundance of natural resources and having a stable social and economic situation. It is the health of the whole system. Rather than telling the ecosystems what to do, rather than controlling their growth, we need to watch ecosystems to see what they do (this used to be the function of natural history), and we need to let them do it (this requires patience and temperance), with a minimum of interference. The way to ecosystem health is letting the ecosystem do most of the choosing and working.

Our response to the ecosystem, being concerned with its health (as ecosystem doctors or nurses perhaps), is not benign neglect or complete anticipatory stewardship, it is participation in the process of the ecosystem as a harmonious system, with mutually restrained conflicts and influences.

The goodness of our lives reflects an imperfect balance of love and selfishness, reason and passion, sensuous materiality and spirituality. We have the responsibility to be healthy, to contribute to the health of our community, and to contribute to the health of natural ecosystems. Good intentions have to be combined with ecological knowledge and ethical behavior for the discipline to be meaningful. Our rules for living together have to be a compassionate participation in the whole planetary process.

Ecosystem Medicine for the Global System

Is the planet really in peril from human activities? Or can it recover if we humans make some changes? The planet has a concept of health, or rather we have the concept, and we can describe it by contrast with an organism, but we have too little experience to try to make the health of the planet a real issue. So, is global medicine even possible?

James Lovelock launched a medical model of Gaia (in his definitive book *Healing the Planet*). He suggested that the earth can be considered a patient that humans can address with some medical training. A few planetary diseases are evident now, such as greenhouse fever, ozone loss, or acid rain 'indigestion.' And there may be potentially fatal diseases at early stages, such as nuclear winter or extinction spasms. As a patient, Gaia is less threatening than a ruthless pagan goddess. Lovelock suggests a practical medicine for the planet, which grows from guesses and empiricism to practical solutions and good hygiene. Of course, this is what ecosystem medicine is—the application to ecosystems, the special organs of Gaia.

Medicine is the identification of problems, such as stress; and, we are aware that humans put stress on global systems. We need to monitor the health of the system, checking temperature and pressure, breathing, biochemical tests and biopsies from ecosystems. We need to increase our understanding of decay and healing. Medicine has shifted responsibility from individual, to be healthy, to doctors and medicines, after decades of dramatic success with some medicines and treatments. This way shifts the burden to the intervener. The way out is avoidance, or awareness of long-term restructuring. Gaian medicine returns the responsibility with the planet, not with an incomplete flawed human management.

This form of global thinking is a critical correction for the machine world view, which traps us in a narrow atomic, context-free, extreme individualism. Some forms of biological thought, such as Neo-Darwinism, have converted Darwin's organic view to a noisy market of manipulative machinery. The metaphor of the machine justifies our indifference to disasters, since nature is considered a lifeless aggregate of atoms.

We can figure out if the planet is healthy—it is relatively stable. But, we are not sure what the planet requires to stay healthy. Does it require asteroid collisions? Does it require the increase in solar radiation? Does it require the ellipse of the planet orbit, with refreshing distances? Lovelock suggested that the galaxy was a giant warehouse containing spare parts needed for life.

Lovelock suggests humans be the stewards, as representatives of the varieties of life, for the planet—not managers, with full responsibility, or masters that make it provide for us. As stewards, we would address species in communities and ecosystems. Gaia as a living planet is a self-regulating system that can correct itself, under most circumstances, and without too much interference. We just need to be pleased that the doctor gets the credit.

Seeing Regional & Global Patterns

Despite complaints and warnings from the overeducated or sensible common observers, about not learning from history, we just *do not learn* from history. We pay no attention to the larger patterns of repetition or convergences of factors that play out over hundreds or thousands of years. Despite annual catastrophes from earth, water or air movements, like earthquakes or hurricanes, we continue to build houses and cities in dangerous places. Despite evidence of planetary wobble from pollution and waste, we continue to pollute and waste on a global scale, creating islands and mountains of plastic, paper and metal.

Perhaps the problem is perceptual; it is sometimes difficult to see large scales from a five-foot platform like the human body. Perhaps it is the limit of the human lifetime; seventy years may not be long enough to internalize changes that take hundreds or thousands of years to be expressed. Perhaps it is a limit of the human mind, that it cannot comprehend the complexity of the trillions of interactions and their consequences. The large, long, slow, invisible trends of evolution are hard to comprehend or adapt to with our limits.

Feedback is also an important component of behavior. But, ecological or evolutionary feedback is slow and small. We can make technological feedback visible. Buildings that have daily readouts or visible monitors might influence people to turn down the heat or turn off lights, to save energy or reduce waste. The design includes a system of participation with feedback loops. Human feedback is critical, also. Design needs to be user centered. It needs to be participatory for the residents and designers, with everyone performing some active role. The goals of design have to include the social transformation of consciousness and positive behaviors as well as more efficient and beautiful shapes. Participation encourages enthusiasm for the project. Community and online networks allow observation of various kinds of consequences under different conditions. This is what thought experiments are!

One design solution for buildings is density, as envisioned by Soleri and others, that is, moving things closer together. The entire city needs to be considered as a design project, vertical and horizontal, below ground and above, self-sufficient or highly connected with trade. Specific buildings need to mimic local landforms or vegetation. They need to perform natural functions, such as damping the wind, holding water, and preventing erosion. The native vegetation surrounding needs to be restored so it can function as wind barriers or shades. Rainwater collection systems, permeable paving, and appropriate structures have to be considered. Design needs to address ways to move people outdoors. It can do this with in-outdoor transition zones and areas for outdoor activities.

All of this basically has to do with health. Health is the harmony of conflicting elements, such as disease, joy, selfishness, and altruism. Ecosystem health envelops the health of species and individuals, human and other. It is not enough simply to reduce disease and violence. Health has to reflect the environment. It has to do with aesthetics, and that includes all the senses—humans need music as much as beautiful forms; they need stimulating smells and tastes, as well as clean water and clean air. Sounds and smells can complement and enhance any building or area, whether the sounds are symphonies or leaves rustling.

True ecological stability requires integration with regional and global cycles. It may require tradeoffs of wilderness and waste sites. We have to consider the patterns of locations of cities around the planet, as well as of the global travel corridors for ships and airplanes, as well as wheeled vehicles, especially buses and trucks. These corridors are shared with animal and plant movements and must be designed to minimize conflict with them. The designs of other of our constructions, such as dams and canals, have to consider impacts on the movement of water as well as the habitats and courses of fish. Human patterns have to respect other larger patterns that have been forming over millions of years. We have the knowledge or understanding to do that. We have the ability to apply that now.

Sofia Echo Series 2000—

Population: How Many People is Best for Bulgaria?

Bulgaria is said to have a 'declining' population. Many politicians and economists are calling for rapid growth. Perhaps because in the past rapid growth has increased prosperity for some people. But, is rapid growth the best strategy for Bulgaria? Imagine that the population has grown to 16 million people—every place is twice as crowded. Has the number of good jobs increased or are all these new people unemployed? Where do they live? Some things do not increase: amounts of land, air, and water are still the same. More people need more land for buildings, factories, farms, and roads. That means fewer places for wilderness, recreation, autos, or buses. High populations in many countries, with many different living standards, are only maintained through the constant takeover of natural habitats for farm land, or through the drawdown of fossil fuels, and by economically cheating the poor and powerless. Eventually these things get corrected by war or adjustments—usually all unpleasant.

What would happen if the number of people keeps decreasing? At some point, maybe half a million, it might be hard to keep the cultural values of the people active and vital. It might be hard to maintain many industrial activities. At very low numbers (5,000-50,000), there might be problems with fertility or social cohesion. The population has been low before, during Thracian, Roman, or medieval times, but it was rarely thought to be too low.

Rather than just let the population grow or shrink, we could try to figure out what a healthy population should be, by relating it to the carrying capacity of the land. The carrying capacity is that number of people who can be supported on a long-term basis, using local renewable and nonrenewable resources for all needs, including clothing, shelter, transportation, information generation, and aesthetic satisfaction. This capacity can be increased or decreased by using different kinds of agriculture or different kinds of technology. For instance, technology can give higher yield crops, but also it hurt crops with unforeseen side-effects (poor uses of pesticides, for example). Furthermore, the capacity decreases as the use of energy and resources per person increases.

The ecologist Eugene Odum suggested using land area as a measure of human carrying capacity. The minimum area requirement per person are is 2.02 hectares to provide all needs sustainably (for the state of Georgia in the US). This number includes land for natural areas, as well for producing food and siting communities and road networks.

Bulgaria and Georgia are roughly comparable in productivity. The size of Bulgaria is 11,091,200 hectares, so if we divide by 2.02 hectares, we get a maximum population of 5,490,693, quite less than the current number. An optimum population might be less than 3 million. Any size human population can disrupt natural cycles and environments, however, so population numbers are only the beginning.

All living beings adapt to and change with each other over time. They change the climate and environment, also. Wild populations usually exist at far less than a maximum number. They are usually limited by the productivity of ecosystems (diseases and predators keep the population healthy and balanced). We humans need a similar flexibility to adapt our populations to changes in climate and the productivity of the land. That is why a smaller population may be better ecologically.

Instead of treating a declining population as a problem, why not consider it an advantage? With ecological planning, Bulgaria could become the first balanced nation on earth, by linking its population to its carrying capacity and its wild environment. Countries such as China, United States and The Netherlands will have to face these limits soon, but Bulgaria could become a good model for them to follow. Let's make Bulgaria the leader.

What Shall We Use Computers For?

Bulgarians are buying computers. Computers are powerful tools. They can process immense quantities of data and solve incredibly complex equations, reducing the time for answers from life-times to minutes. Businesses can be run from a laptop computer, using sophisticated programs. Computer simulations are cost-effective and literally life-saving in fields like medicine. In ecology, computer models can be used to estimate the effects on hydrological cycles from cutting a forest. The only cost of a simulation is time or dollars. This is not true in real systems, where acid rain kills fish or trees.

Computers are now allowing us to live in an amazing hyperworld of the Web, where communications propagate like electronic wildfire through the system, where people are brought together by the computer. This wild place could make our lives come alive with beautiful graphics and vital exchanges. Businesses and schools are excited. Teachers promote the strategic roles of computing. Salespeople tell everyone that computers save time, increase efficiency, and solve problems. Sometimes they do. So, people are pressured into buying and using computers.

There is no doubt that computers have brought about many good changes, as have the phone or automobile. But the phone and auto had undesirable effects; the invasion of privacy and pollution, for example. What are the undesirable effects of computers? Loss of freedom? Structured education? Computers are being used unthinkingly for any application. But, as the radio was not very good for comprehensive communication, computers are often not very good for creative construction or nonlinear exploration. The computer is becoming a dominant all-purpose machine, but it may not be a solution for the difficulties of modern civilization. In fact, excessive reliance on any machine may have distinct disadvantages for us.

The use of computers may have long-term psychological effects that should be considered. If we consider machines, like computers, as "energy slaves," then we each use the equivalent of ten to fifty slaves per day. We slave owners love the privilege (I do). And the exploitation of inanimate slaves is more easily justified. Perhaps now, the culture of arts and sciences is not possible without machine slaves. But, at some point, slavery corrupts the owners, making them physically or mentally softer. What is the exchange for summoning these information slaves so easily? The failure to develop intellectual ingenuity? Loss of imagination? Lack of trust in intuition?

Consider an earlier adaptation that helped humanity. Knives permitted hunting larger game animals, but the long-term anatomical result was partial degeneration of the human jaw (how many of you have wisdom teeth?). An adapted species is more vulnerable to accidents, since external adaptations, like pacemakers or artificial kidneys, become required for the health and maintenance of civilization. Perhaps writing changed human memory. Perhaps computers will change our thought processes.

Computers allow us to keep track of inventory without memory and to add without effort. But, when computers fail, people have more difficulty adding up their grocery bills or keeping track of all the widgets made. Computers successfully augment our abilities, but we must take care that they do not allow the abilities to degenerate. Computers are valuable, but we must not forget what functions the computer are assisting—computers should not displace the skills themselves. Education should include a core of mathematics; poetry and narratives should still be memorized, as well as written. Our lives should include working in the fields or resting under trees. Ultimately, we want to live harmoniously on earth, with the wealth of other living beings. Computers can be an important part of our lives, but not necessarily a crucial one.

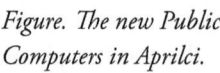
Figure. The new Public Computers in Aprilci.

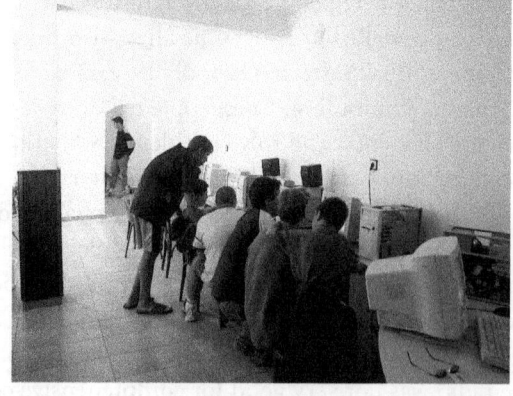

Are There Too Many Roads in Bulgaria?

Bulgaria has many roads, although politicians, as well as business people and drivers, are asking for more, wider, smoother, and direct roads. Cities want better roads leading to their centers. Forestry managers want more roads into forests. Resort areas want more roads leading to the sea or mountains. Economically roads can stimulate income, at first. But, there are other ways of looking at roads.

We can look at roads from an ecological perspective. As a science, ecology describes the interrelationships of organisms and environments, that is, the experience of living together in the biosphere. But, ecology is also a way of "seeing" that human beings are participants in nature, as part of the food chain, for example. People, like most mammals, use roads (or paths) to get to a place, to get supplies, to visit others, or just to look around.

This is a fine use of our technology. It allows us to increase our horizons and better our lives. Better roads make traveling more efficient. There is less waste of oil and gas, and less wear on vehicles and their occupants. But, every technological innovation in vehicles either requires or makes roads. Roads have effects that go far beyond the movement of people on them.

New roads lead more people to new places, thus changing the characteristics that often make those places attractive (e.g., being off the road). Roads increase the flow of things between points. But, too much flow (of matter, energy or form) can destroy biological relationships and diversity. Roads are a major force in fragmenting the habitats of plants and animals. Many animals cannot live near roads or noise or human activity. Many animals and plants need large areas to roam and roads cut into their areas (although, highway routes and underpasses can be modified; for instance Britain builds underpasses for frogs). Roads directly affect natural and human communities in many ways, causing:
- Changes in populations of animals or plants that cannot cross them (isolation)
- The spread of organisms that use roads to colonize new areas with plant or insect or animal pests (that is, things that are out of place)
- Problems with erosion
- Problems with spreading trash (other things out of place)
- Changes in hydrology and wetlands
- Changes in social circumstances. For instance, private cars changed public morality in America. Many kinds of crime are increased, for instance, bank robberies, if there are fast roads nearby with easy access.
- Changes in economies, as new roads bypass old routes.

Many countries answered the demands of their citizens, business people, and politicians (and ignored the environment) by building bigger, faster roads. Then as people crowded on the roads, they get more crowded and slower, and the demand for bigger roads rose again. Many countries have found that building more and larger roads does not solve the problems of congestion, accidents, and danger. These problems have a lot to do with the kind of transportation on the roads, that is, cars, trucks or buses.

Maybe Bulgaria should have a new autobahn and maybe all the roads should be upgraded. But, it would be better if it were part of a plan for the entire country that considered population movement, the needs of all the people, and the best forms of transportation. Buses and trains are far more efficient than private cars, and many countries are rebuilding their train and bus routes, from Brazil to France and Japan. By concentrating on a good public transportation system, and limiting the influence of private cars, Bulgaria could become a good model for them to follow.

Is the United Nations Relevant to Bulgaria?

After the war ended in 1918, there was a popular vision of One World, without borders or barriers, created through reason. Our many individual national attempts at social improvements, however, have proceeded without adequate reason, without order, without sufficient perspective, without adequate confidence, without a comprehensive plan, and without a great dream. Historically, the creation of states has not been through reason. The boundaries of almost all countries, from China, the USA and Mexico, to Germany, France, Italy, Greece and Turkey, are fairly arbitrary, being based on the results of military struggles in the past. These and many other countries were united by force. Our efforts to provide the infrastructure for our global civilization have been guided by anonymous builders, mediocre designers, minimalist engineers, rapacious financiers, and corrupt politicians. Yet, the notion of a world government seems to satisfy a basic craving for unity and order. And, an implicit world system is evolving slowly through economics and science. But, a new global order is necessary to govern the system. At the current stage of international relations, there seems to be no agreeable path toward such a world order. The partial adoption of international institutions is insufficient for a world order, especially if those bodies are only advisory.

The United Nations (UN) is the only existing body with the machinery for constructing a world order. However, as long as ecological and political problems are addressed in a framework of nationalism and military power, the UN is treated as peripheral and relatively impotent. Furthermore, as it is structured, the UN is not capable of handling the responsibility for world order. For example, restricting membership in the security council to powers with nuclear arsenals, or using the veto principle, indicate problems. Furthermore, even when the UN does make good recommendations, it does not have the power to coerce any nation to follow them. Rather than replace the UN, countries must revise it. The UN has been a half-hearted investment, but it has historical appeal and wide support. It is a nascent global order, but it must have new structures and new functions, new powers and new responsibilities. Two are especially important now.

The first is protecting diversity. The UN must have real explicit responsibilities, such as the protection of biological diversity for the planet, as well as of human cultural diversity. Around 1900, there were over 1000 different human cultures and over 3000 different languages (roughly equivalent to the number of natural biogeographical provinces and habitats on earth). Cultural diversity is necessary to protect biological diversity, the ecological wealth of each country. If we diminish variety in the natural world, we debase its stability and wholeness. If we wish to advance human civilization, we must preserve and promote variety. New countries, such as Kiribati, Liechtenstein, Marshall Islands, Monaco, and Vanuatu, have joined the UN recently. Many more countries want independence based on their cultural and linguistic uniqueness. Perhaps many of the borders drawn by violence, in Bulgaria, Albania, Croatia, Turkey, Iraq, and Macedonia, among others, could be redrawn peacefully e.g., give the Kurds their own nation).

The second is protecting security. The UN must also be given real powers for protecting countries and for policing international terrorism. Most military power, except for local police or national guard, could be turned over to the United Nations (nuclear disarmament itself could be accomplished relatively quickly with complete international support). Wars are being fought without an international referee with power or respect. Violence will continue (regardless of how well justified—and justification these days has a very dissolute and tangled history), although the cycle of attack, hatred, and revenge can be broken by an international body composed of representatives of all peoples. International terrorism, or criminal actions, such as those recently in Iraq, Afghanistan, Macedonia, and the USA, require an international response, with an international police force, an international justice system, and an international form of punishment, so that all the countries of the world have a say and a stake in the peace process.

Having the UN address international issues, as a confederacy of concerned neighbors, would do much to diffuse the polarity of one country trying to be world leader and peacemaker (especially when the USA, for instance, is viewed as the power behind many thrones that exist only to protect US economic interests). The UN could provide support—food, health aid, engineering —to a country such as Afghanistan and it would not be resented as a gift of rich people wanting a market or an ally. It would just be the act of good neighbors.

Does Hunting Have a Place in Bulgaria?

As phylogenetic omnivores, humans have a long tradition of eating and using animals. Many early cultures revolved around hunting as a way of life. These cultures had traditions that revered their prey and many taboos to limit taking prey animals; many prime game animals were taboo altogether. Partly as a result of successful competition, human groups expanded and came to enter and dominate virtually every ecosystem, such that scale, greed, and ignorance resulted in extermination, overgrazing, overcutting, and wanton destruction.

Recently, hunting in Bulgaria was necessary for people to supplement their gardens and farms. But, the hunting was regulated by the national hunting society, which controlled or limited the number of animals taken. Now, since the transition to a market economy, and with the dissolution of the national hunting society, hunting has become more chaotic, and more numbers of many game animals, from red deer to wild swine, and trophy animals, such a wolves and bears, have been killed.

Many magazine articles glorify hunting. Understandably, these articles do not address all of the issues associated with hunting or with animals as "natural resources." These articles address a limited perspective of hunting, that associated with the individual pleasure at stalking animals and using technological advantages to kill them. There are other dimensions and many larger questions unasked. One is the shear number of hunters, including sport hunters, hunting tourists and poachers. Another is the myths based on early human traditions. These can be dispelled.

1. "Hunting is necessary to keep game species in check and healthy, even if original predators are reintroduced." No, the population rarely increases beyond the carrying capacity before it stabilizes. Food is always the final check. Predators stabilize a herd at a lower number and keep it on the move, which improves its health.

2. "Hunting is necessary to protect human economic interests, especially crops, cows, sheep and chickens." Not so. Usually preventive measures are enough; good fences, good dogs, and enough attention by shepherds and farmers.

3. "Hunters take the place of natural predators to maintain the balance of nature." Natural predators are more efficient and cost-effective. Why create an imbalance by killing predators and then try to take their place, mostly unsuccessfully? Hunting as a system is very unnatural, in its timing, its targets (usually the largest), and its scale.

4. "Hunting has a minimal impact on game species and almost none on nongame ones." Many habitats have been artificially skewed towards game species. Hunters often shoot prime specimens, which is the opposite

tact taken by predators, who take the sick and old from herds. Also, many animals, especially rare species—wolf, lynx, marten, panther—cannot survive hunting.

5. "Sport hunters are concerned about their image, and only a small percentage are slobs." Hunting takes less agility and conditioning than other "sports." The lack is made up for by motorized transport (snowmobiles, trucks, dirt-bikes) and by powerful, sighted weapons.

6. "Hunters are just following nature's law of kill or be killed." It's an old law, never properly understood and repealed by new knowledge. The new law is cooperation, with limited competition.

7. "Hunting is okay because animals have few feelings and emotions and feel little pain." New evidence shows differently; even fish have well-developed feelings. Let's stop pretending.

In practice, sport hunting demonstrates an ignorance of human nature, of animal nature (more than just a series of tracks leading to a target), of wildlife management, and of interactions in ecosystems. It demonstrates an insensitivity to the feelings and goals of animals, and it demonstrates an abuse of power. Animals are threatened with loss of habitat and deteriorating ecosystems, from take-over, development, human overpopulation, and pollution, as well as from hunting. Hunting puts pressure on animal populations without noticeable benefit. Hunting is a minor part of the total of human outdoor activities, numerically or economically. Far more people walk outside, take pictures, ski, or camp. Perhaps we should just eliminate hunting. After all, humans with brains and spears were more than a match for mammoths. And humans with super technological informational systems, weapons, and vehicles have shown that they are overmatched for virtually any species, even without bringing those brains into action.

Is Bulgaria Wealthy?

Recently I was asked what I thought about poverty in Bulgaria—the lack of cash that makes Bulgarians poor. I did not know what to say. From what I see, Bulgaria is rich. Perhaps we misunderstood each other's ideas of wealth.

Economics has always been concerned with measuring wealth. Wealth once meant tangible things, such as land, ships, and houses. Later, it was measured by labor and production. Now, it has come to mean negotiable symbols such as cash and stocks. Information is now considered wealth.

The first economies depended on their own food and minerals. The mass economy rose after the industrial revolution, when networks of governments and institutions were created to hunt and acquire vast quantities of material in order to manufacture products that could be sold at profit. This economy is being altered by increasing populations and by increasing difficulties in finding cheap resources. Some of our perceived wealth and assets are disappearing in the process. But, symbols such as cash are considered new forms of wealth. And with computers, information itself is valued as the ultimate resource and source of wealth.

Information is apparently boundless. Yet it can be manipulated. It is information that defines the use of resources by people. For example, hydrogen is worthless unless technologies exist to transmute it to helium and manage the released energy.

Land and resources are considered less important. But information without "form" is nothing. Information lets us use resources and land more efficiently. Land and resources still are part of the basis of wealth (a material dimension)—as many native peoples have found out when coal or pharmaceutical plants were discovered on their lands.

The narrow definition of wealth (as just one thing, resources or information) means that it can be increased only by producing a bigger supply of goods or reducing the demand for goods. Wealth is defined as supply divided by demand. If supply is limited then wealth can only be increased two ways: reduce the expectations of individuals (smaller pieces of the pie) or reduce the number of individuals (fewer larger pieces of the pie). Supply may be mostly material things—but not status, for instance—while demand has the more psychological dimension. Wealth has a psychological dimension. This dimension is not limited by strict logic. Wealth can therefore be expanded without being limited to supply or demand for materials.

The assessment of personal or cultural wealth, for instance, is mostly psychological; wealth may be measured by how many valuables one has, which may be physical, like feathers or salmon or gold, or by how by much status one has, which may be behavioral, as when enjoying deference or a good reputation.

Rich sensory experiences can be derived from direct contact with nature. But economists rarely mention these values. Light, wind, dirt, plants, birds, all act on us—but not with the meaning of crops or vehicles, which is for their utility—they just are. People do not live without these things. They are valuable to us.

Until now, economics has required growth to increase wealth. Growth has been a substitute for equality; it seems to be necessary to avoid revolt—even after 400 years, growth has not brought wealth or equality to most people. Ecologically, the goal of economics should be mature development, not growth. Development means the introduction of an innovation. Economic development will still require technology and new forms.

A mature economy would be like an animal or plant, or a mature ecosystem, like an old beech forest. In its early stages, a beech tree can still be stable. Growth in trees can delay the onset of senility by ridding it of waste products in more diluted form. However, too much growth produces a strain on tissues and early decay. Later stability must result from limits and metabolism. When it reaches a size that fits its genetic and environmental limits, it is mature. It continues to change, but that is development of new relationships and forms.

It seems that the wealth of a country is a function of its physical attributes (resources) and its culture (application of information by people). In fact, the attributes are only possibilities until appropriate cultural perceptions and technologies exist. The inclusion of nature as a source requires an ecological dimension to wealth.

Bulgaria has wild lands with all kinds of unique wildlife and plants. Bulgaria has fertile fields that could grow all the food that her people need. These things are an important basis of true wealth. Bulgaria also has an educated and exceptional people, who have the knowledge and information to make sure that some of the natural wealth is transformed to human wealth. By this most advanced definition of wealth, Bulgarians are indeed wealthy.

Is Capitalism Better than Communism?

Formal development is more concerned with an assembly-line model—simple, isolated, efficient, and easy to maintain. But, we become remote from, and indifferent to, the system that supports us. We acquire unrealistic images of the world and harmful values and then make bad decisions based upon them. We have not developed qualitative indicators of ecological health or quantitative measures of social health.

Traditional economics has a three-capital model of human wealth. Land, labor, and manufactures. Clearly, this is not adequate; the definition of each kind of capital needs to be expanded. For instance, land is the entire ecological system, complete with other species and biogeochemical cycles, preserves, as well as agricultural areas, resources, and artificial modifications like dams. And labor depends on the traditional capital of a culture, the beliefs and myths and rules for behaving, the institutions. Manufactures themselves depend on culture and land (resources), as well as on culturally appropriate technology.

The distribution of symbolic and real wealth is very inequitable as the result of historical trends, old economic rules, and cultural confusion. The inequity could be reduced in Bulgaria with caps on top salaries, a minimum income (or negative tax), and a simplified 10 percent tax on income. The services provided to society by some groups, football players, lawyers and politicians, for instance, are paid for disproportionately well compared to those of others, such as teachers or nurses or policemen. Recognizing that the market is not the best judge of social value, local government could set salary ranges.

We make trade-offs in social systems without assigning dollar values. We could do that with ecosystems. The valuation could be scientific or be based on human labor. New values of natural resources are being recognized now by economists, such as option value, that is, reserving a resource for future use or existence value or paying for something to exist that will not be used.

One thing business can do is put a price on nature. But, let us make it a real price, reflecting the real cost of replacement. Let us base the cost on human labor and technology, so that 1 gallon of oil is worth a million dollars, as Buckminster Fuller once calculated. Let us make all those prices high, too high rather than too low.

The formal study of how people use their surroundings is economics (from the Greek words meaning "law of the house"—house is used as a metaphor for human society and nature). The word has come to mean the management of resources to supply human needs.

Modern economics is defined, by Chisholm and McCarty, as "the way people make their living." Economics attempts to address the 18th-century concerns of Adam Smith by using a scientific method to collect and interpret information—in fact, economics considers itself a science like physics, chemistry, or biology. Smith, whose first book was on morals (in fact, he considered economics to be a branch of moral philosophy), had noticed the trend in England away from mercantilism (where the central government regulated the output of goods for trade for gold) towards a free-choice economy, where people would decide what to make and how much to sell it for. Smith thought that a free market of independent buyers and sellers would let the entire community prosper as if an "invisible hand" (metaphor alert) were guiding it. That is, competition would increase public well-being, or to say it in a different way, self-interest is linked with common interest (or "What's good for General Motors is good for the USA"). Many countries in the 18th and 19-century, including the United States, adopted the free market system, with the result that the citizens did acquire more symbolic wealth.

Figure. The new Capitalism pyramid (poster).

Is Community Better Than Capitalism?

Bulgaria has accepted much of the help offered by capitalist countries, and seems eager to replace the old communist command model with the old capitalist market-oriented economy. Is this a good idea? Will capitalism automatically improve things? Is there a third alternative, after communism and capitalism, that Bulgaria should use?

The communist model provided many things for many the people, while maintaining a large war and research establishment. The philosopher Karl Marx thought that socialism was inevitable, that public ownership of the means of production would provide equality and social security for all people. But, in practice the distribution of wealth was very inequitable, as the result of historical trends, old economic rules, and cultural confusion. Some people were more equal than others.

As a result of central planning, the patterns of life under the communist model were pressured to be uniform and efficient. Yet, the strengths of this kind of economy—especially the planning of production and the control of resources—were not admitted. Instead, military competition ruined these command economies and socialism is considered a failure.

Capitalism developed as a 'game' for dealing with excess. It ignored inequities in large societies, but people accepted those based on the promise of them acquiring more. Capitalist or market economies are exploiting the resources, such as the forests of Siberia, of failed economies, while at the same time trying to rehabilitate other resources, such as the forests of East Germany and Poland. This contradictory behavior is due to its own economic myths. Capitalism is based on erroneous assumptions about nature and culture, for instance that "nature is a resource to be exploited" or " anything can be substituted for anything else." Marx thought that free enterprise was the most efficient and dynamic system, but he—correctly— identified its basic flaw as the accumulation of wealth without the capability of using it wisely. Misery has increased as fast as wealth.

The capitalist model has proved to be a limited, flawed and self-serving system that has many negative effects, such as unemployment, poverty, homelessness, environmental ruin, and wild inequity. Americans are criticized by the French, not unreasonably, for having a "frivolous" culture based on "savage" capitalism. Capitalism also increases the pressure for uniformity. The patterns of life have become the products of market forces and stylish transportation operating in a sterile abstract order. It is as sterile and poor as the communist model!

The choice is not necessarily between communism or capitalism, two old flawed systems. There are other models for economies, such as the Scandinavian model, or the Eco-models (described by Herman Daly and John B. Cobb, Jr.), or community anarchism (described by Paul Goodman). These models argue for strong communities first.

We do not need to surrender to fast, giant, national economies. We need to shift power to local communities, through self-reliance and participation. Community economics has many advantages over communism and capitalism. For instance, community economics allows barter and other informal exchanges. It also offers more services far cheaper than a formal modern corporate system. A community protects individual freedoms, guards regional culture (values and identity), and holds groups accountable for their use of power. In communities, people can decide to be conservatively sustainable or to grow and gamble on innovation. Communities can have different economic attitudes, paces, and goals. A community that is balanced and flexible, in tune with natural cycles, based on traditional values—in which industrial production is limited to appropriate goods—can position the community for a long, sustainable future in a healthy environment.

Bulgaria has numerous small businesses, e.g., clothing, food, and hardware stores, automotive repair, and lumber yards and construction companies. But, it also has large business that allow money to flow to foreign countries. Certainly, the country could implement ways to keep money circulating in local communities; sometimes this can be done by simply buying locally, but more ambitious solutions, such as local barter "bucks," are also possible. Bulgaria could still differentiate its unique products, within a regional partnership, such as the European Union. This would benefit Bulgaria, as well as the larger regional community.

Stuff Out of Place: Where Should Trash Go in Bulgaria?

Recently a Bulgarian colleague went to a tourism conference in Romania. I asked him what impressed him the most about his trip: He said there was no trash! None on the roads or rivers, none on trails or at rest sites. I asked if trash was mentioned at the tourism conference; he said that it was not.

Throughout our history we humans have always left behind things that we could not use. Some of these kitchen middens are very valuable to archaeologists; they tell us a lot about the lives of our ancestors. Other of our leavings have been of great value to animals, such as rats or birds, who have learned to live with us and benefit from our messes and excesses.

As we became more numerous and more technologically advanced, we made things that rats and birds could not use, and that we could not reuse easily. These things formed concentrated pockets of trash on the landscape (trashscapes?). When these places became too ugly or dangerous (causing pollution and illness), we started to clean them up. We noticed that places that were ugly and dangerous were not popular vacation destinations or desirable places to live (although it seems that the poor cannot escape them as easily as the rich).

I described to my colleague my first day in Bulgaria. I got off the bus next to a dumpster full of rotting fish, then walked along a trash-lined street to the place where I met my hosts. I asked my colleague why there seemed to be so much trash in Bulgaria. He said that under the previous rule everyone was required to work for half a day once a week cleaning up trash— since the people were forced to clean up, they got in the habit of littering while walking, riding, or driving, knowing that they would have to clean it up later. After the transition to a new political and economic system, they kept their habits of littering, but no longer had to work to clean it up.

Trash is merely a resource that is out of place. Many resources are out of place in Bulgaria. There are several possible solutions. We could just accept it (and help future archaeologists). We could develop ways to keep the resources in circulation, by recycling or forming industrial partnerships among companies that could use the waste of other companies. Failing that, we could collect and recycle many kinds of trash. And, failing that, we could store it in landfills until a way is found to deal with it.

Trash is a major problem for any country with ambitions to be a tourist destination. Many countries, such as Switzerland and Romania, have discovered that it pays to keep trash out of the rivers, forests, and streets. People want to visit nice clean places in other countries, either pristine nature or culturally significant sites. Bulgaria has many wonderful places to visit, places of which that people can be proud. But, throwing trash in them is not the action of people who are proud of their country or who want other people to visit it (and pay to see it and to stay there for more than a day—unless of course, it is unique because it is the trashiest place of all).

All of the homes I have visited here are scrupulously clean; even the land in front of the homes and the parts of sidewalks and roads are kept clean by the families. The other parts of the roads and wild nature, that are held in common, need to be treated in the same respectful way. The simplest thing would be to ask people not to throw things along the roads or trails. But, perhaps the State and Cities could ask people to volunteer some time to collect all the trash that has been covering the beautiful places. Many other countries have found that this is simply in their own self-interest.

Note: Trash in Aprilci and many cities was left for the gypsies to sort; the gypsies then burned the rest in the dumpsters, so local residents could refill them. In Troyan, Dupnitsa and other cities, trash was collected and taken to a dump by the river, where it was sorted and burned by gypsies, then later bulldozed over the hill, at least until 2002.

Are There Too Many Pets Here in Bulgaria?

Domestic animals, especially pets, are truly loved by Bulgarians, but there seem to be large numbers of them roaming the streets. Many cities are trying to solve this situation by, for instance, killing all the stray dogs in a city. Maybe this will work for a while, but we should ask ourselves, what the role of animals is in our lives, and what effects do they have on us, and on artificial or natural ecosystems.

Animals have been part of our lives since the first humans. Throughout our history, humans have killed animals for food and clothing. Comparatively recently, animals and then plants were domesticated. Many animals were attracted to human settlements, often because of leftover food, but perhaps because young animals like to play with young humans. Wolves were domesticated and bred into dogs. Cats were fed and tolerated. Fish and birds and reptiles have been taken into human households.

As human technologies developed, our relationships with animals changed. Hunting, grazing and agriculture produced large ecological changes in nature. Early domestic animals were revered, but wild animals became competitors or nuisances. Lately, animals have been treated as commodities processed in factories. Emphasis on economic efficiency has resulted in utility hens in batteries and milk production from cow factories. Wildlife is considered useless. Hunting activities persist, but mainly as recreation.

Both domestic animals and pets have important roles in our lives. More than just useful for eating leftover food, pets are social companions, and unpaid therapists, especially for old people, the sick, the young, and the lonely. They are more playful and forgiving than many human companions, that is, they accept us for who we are, without demanding that we change or improve things for them, beyond feeding them and giving them a place to sleep. They also have educational value to us; they can teach us how to relate to nonhuman nature and how to develop as fully human beings.

Caring for pets teaches us responsibility, but caring may also be an adjustment to modern life, a grasping for ecological connections in a technological world. Our attitude towards pets is often characterized by an incomplete economics—We do not think of all the costs associated with pets—and by a faulty ecology—We must learn that pets are still connected to a wider world.

There are ecological and humanitarian problems with pets. Pets consume a considerable percentage of our grains and meats, as much as significant numbers of people. Dog excrement carries diseases. Now that Paris has an anti-excrement law, perhaps the cities of Bulgaria should consider one. Not just for hygiene and aesthetics, but also for economic reasons, and not just the costs of cleaning clothing and footwear, but the costs of lost tourists. Noisy dogs keep people from concentrating, resting or sleeping. Dangerous dogs not only hurt people, but can infect them with diseases and interfere with wildlife, killing birds and deer in an area.

And, we are the problem for many pets and animals. We abuse many pets. Millions of unwanted pets are put to death every year (12 million in 1984 in the United States alone); millions are used in experiments; hundreds of thousands are shot for food or pleasure. People are ignorant of these facts or detached from the consequences of their personal actions. Our prevalent attitudes towards pets are usually human-centered or negative, either the affection for or avoidance of individual animals. Traditional humane movements have tried to teach us that pets are entitled to fair treatment when under control.

What should we do? Obviously, we should not have more pets than we can care for and be responsible for. Having a pet is more complex than just having fun with them and feeding them. It entails making sure we have time to care for them, walking them and cleaning up after them, making sure that we can keep them when they are *not* as small and cuddly, making sure that they do not harm others, either people or domestic and wild animals. When we fulfill our commitments to pets, then there should not be as much of a problem with unwanted and dangerous pets.

Are Pets Needed? Should We Destroy Them?

I went to the local hospital once to pick up some rifle bullets to kill wild pigs in the mountains overlooking the spa. I had been assigned to do that by the national park. As I approached the hospital on foot—the only way to travel!—I saw what looked to be a dark living carpet on the wide concrete stairs. It undulated in the sun, waves rising and falling in some strange pattern. As I walked closer the particles resolved themselves into dogs, mostly small and young. They parted as I reached the stairs, possibly from habit.

I asked about them from the patient who had the bullets under his hospital bed. I didn't ask if he had been shot. He gave me the bullets and we spread them out on his bed. Most of them were for shotgun or luger. I took them and thanked him, letting him know the ones I didn't use would be kept for him at the park office. He was happy to contribute. He mentioned that the dogs had been here the past five years. I nodded.

After the bank collapses in 1996, most people freed their dogs and cats, since they could not feed them any longer. The dogs formed packs and stayed in town, living on scraps, garden vegetables and cats. He said they never bothered anyone here. It was a different story in other cities. Near my office they had attacked individuals, usually at night. Before any large National or International Event, the dogs are rounded up and killed, so the dignitaries would not see them. Of course the mafia had a business of skinning them and sending the skins to China for shoe leather; that resulted in corpses floating down the river—the first time I saw one I dove into the river thinking it was a child; four days of bathing removed the stench finally.

Our preservation of animals has unspoken rules. They tend to have to be charismatic, large animals like bears. They are rarely competitors like rats (although wolves can compete with ranchers). They are not human parasites, like fleas or liver flukes. They should be mammals or birds, not clams or fungi. They should be large enough to see, which lets out bacteria.

Pets, and a few wild but caged birds, reptiles, and mammals, need to be considered for fitness into nature. Because of pets, we are less concerned with wild animals and allow them to slip into extinction. Then, too, we maintain our pets out of context; they are fed food from cans, kept indoors or in cages, and taught to defecate neatly. Zoos, aquaria and aviaries keep animals alive out of context, also, out of the wild conditions that shaped them. Denied the possibility of biological meaning, many of the animals go mad. We will never understand why they are the way they are, if we only study them in our homes or their prisons.

Pets only remotely connect us to the wild world. Their value now is as a replacement for wild animals, and often a replacement for emotional entanglements with other human beings. We dress up these animals as humans, and are disappointed when they do not act like humans. On the other hand, we appreciate their unconditional love for us, the recognized alpha animal and often leader of the small pack.

Pets are now significant consumers of expensive meats, beds, clothing, and sophisticated medical care. The expenditure on pets runs in the billions of Euros. And, there are too many pets that suffer in poor conditions. As they have been bred for so many reasons, for size or shape for instance, the number of their illnesses and defects has risen.

Therefore, we have to limit pets, to drastically reduce their numbers. We can start by banning commercial breeders and rogue 'puppy mills.' We can require the current animal shelters to be emptied by adoption. This would be a transition to nations with many fewer pets, and these would be treated as partners. We can require licenses for ownership. People would have to pass an examination to prove they understand the home and dietary needs of their pet, before they can buy them. They would then be responsible for that animal for its lifetime. Many people would not be granted a license. Working animals, such as seeing-eye dogs, would continue to be used and licensed. There would also be limits of the number of pets one person could have—perhaps one. Many pets could be shared in neighborhoods; this kind of sharing has worked in hospitals, nursing and retirement homes. Pets that have gone feral would have to be rounded up and rehabilitated, or killed mercifully, to save many millions of wild birds. (The next day, I declined the chance to shoot pigs, so I could track wolves, which ate pigs.)

Isolate Bulgaria from Global Economics?

Bulgarians feel they have been reduced by other countries, and ignored as being unimportant or underdeveloped. There is some truth to that, of course. Bulgaria has recently aspired to increase its territory to what it has before 950 AD. And, it allied itself with other countries looking for living room, but the wars did not work out favorably. Nevertheless, Bulgaria has immense wealth for her people. Its biological diversity is great and respected; the nation sits at the crossroads of Africa, Asia, and Europe, biologically and politically. The people of Bulgaria are hard working and inventive, in everything from clothing to computers. They are also curious and friendly, which makes visitors feel comfortable and welcome.

Modern global economics has isolated Bulgaria to some extent. And Bulgaria seems isolated from the volatile politics driven by larger nations like China, Germany, the USA, and Russia. But, that is not always undesirable or a problem. Of course, isolation is what allows a culture to develop a strong identity, and work out a reasonable exploitation of its natural environment, which should be the object of intense study and understanding, so that the exploitation fits and is balanced with the uninterrupted processes of renewal of the environment.

With entrance to the European Union, Bulgaria has realized many advantages in security and exchanges. However, before that, Bulgaria was self-sustaining and frugal. Now, Bulgarians have discovered debt, which has allowed them to buy new homes, automobiles, and appliances; small business owners have bought larger businesses like restaurants or automobile dealerships. The number of imports has risen dramatically as has the total national debt, which will require more use of the environment, as well as more manufacturing or tourist dollars. Development will accelerate and people will be bound up in working faster to compete with other nations. It seems this is just the price of globalization.

There are advantages to being able to decouple from a global network, however. One is to avoid cascading crashes from financial bubbles. Perhaps if Bulgaria embraced its isolation and became a model for other nations, more good would come from that. Often limiting access to a nation makes it more desirable. Its resources and products would acquire a higher value. Its parks and cities would seem exclusive and valuable. The nation would be seen as one that knows its own worth and did not want the rush of progress to destroy its advanced structure of cities, parks and wildlife—those things that every Bulgarian can see and visit. There may even be an increase in a National Happiness Index (recommended in an earlier experiment)!

Can We Live with Wolves in Bulgaria?

Wolves started returning to Bulgaria about the same time that other transitions occurred in the human cultures. With the creation of new larger national parks, a good percentage of territory has been set aside for wild nature—much of this is good habitat for wolves. However, many of the old attitudes towards wolves remain, so that there is hunting, poaching and random killing of wolves throughout Bulgaria.

The wolf is one of the largest members of the dog family, domesticated to form a domestic alliance with humans. In spite of this, wolves are feared and persecuted, and now are endangered or extirpated throughout much of their range around the world. Wolves are excellent hunters and prey on large hoofed animals, such as caribou, deer, and elk. Wolves in their original range roamed over all of Bulgaria. As the human population expanded the wolf population diminished or disappeared from many areas, such as the Stara Planina Mountains in central Bulgaria. In the 20th Century, the activities of the government and hunting societies virtually eliminated wolves throughout most of the country. Small wolf populations may have remained in the Pirin and Rhodope areas in the South of Bulgaria. Healthier wolf populations remained in northern Greece and possibly in the western Balkans.

Wolves increased in many areas after the breakdown of the former regime because the governmental control systems with bounties were no longer in place. But then, due to a dramatic decrease of ungulate populations, which is most likely the result of uncontrolled poaching, wolf populations are becoming endangered. Poachers may have almost completely wiped out ungulate populations in some areas. This may be true also in Romania and Slovakia. As a result of decreasing food resources, wolves turn to livestock as easy prey, although since the political transition to democracy, livestock numbers have also been decreasing. This has the effect of increasing hunting pressure on the wolves.

For a long period of time, Bulgaria has been influenced by the traditional idea that large predators, mainly the wolf, are the enemy of game and livestock. In the last several years, the opinion became somewhat more moderate and wildlife managers accepted the role of the wolf in ecosystem dynamics, but hunters and game keepers in the field do not want to accept the existence of a large population of wolves. The traditional myths and folklore, not to mention the evidence of wolf predation—wolves also act as efficient scavengers of large mammals that die from starvation or injury— are hard to overcome. Sick, injured, young, or aged animals that lag behind their herds make easy targets for wolves. Wolves help strengthen herds by killing such animals. Old or unhealthy animals can be a burden on a herd. By eliminating such animals, wolves perform an important natural function in wild ecosystems.

Public attitudes towards wolves, especially in rural areas, are still very negative. Many people fear wolves, because of their reputation in myth and folklore, and also because they believe wolves attack human beings (the howl can be frightening). But, wolves avoid people as much as possible. Many people hate wolves because they kill domestic animals, such as sheep and cows, but wolves prefer to eat deer or elk. Wolves are blamed for many livestock losses, which may result from poor health, accidents, or feral dogs. In fact, poaching causes more losses of game animals than wolves. Many hunters dislike wolves because they kill game animals, such as elk and deer, but wolves usually only take sick or young individuals and not healthy trophy animals.

Wolves can coexist with human populations in Bulgaria. As people are educated (especially through the efforts of the *Balkani Wildlife Society* and their programs for schools and shepherds) about the ecological significance and value of wolves, they can accept wolves as part of dynamic areas that surround our humanized cities and fields. There are many habitats set aside for wild nature. But, a rational policy needs to be established to allow the habitat to be protected.

The Limits of the Planet

Humanity is using vast amounts of physical resources for its industrial goods. Vast acreages of land are being converted from forest and wildness into agricultural fields and terraces. Energy is being captured from fossil fuels to drive agricultural productivity and economic development; at the rates being used, additional reserves from ocean drilling and fracking deep gas pockets would only extend our industrial machine for months, not even years, although for an election cycle the numbers appear worthwhile in terms of financial and environmental costs.

As the global economy races to low wages and cheap production, it starts to run into the limits of the economy itself—there is no room to expand further. Furthermore, it will reach the physical limits of the planet. Ecosystems can only be so simplified before collapsing; they can only absorb so much extra energy and elements like nitrogen and phosphorus before simplifying; there is only so much copper that can be mined. Eventually, the global economy will not be able to expand; it will have to conform to the scale of the planet. It will have to deal with the exhaustion of critical resources and the extinctions of critical species. There may not be any room to maneuver. There might be some fertile lands for crops, or even available rooftops and sides of buildings, but they may have to be supplemented by food grown in vats and flavored artificially.

We are not sure of the ecological limits of the planet. We do not know what a maximum or optimum take is for wild food species. In many cases, we do not know what our exploitation, even if it is reasonable, does to the food web. We have not measured our inputs into wild systems, whether it is disruptive amounts of energy, or excess elements such as nitrogen or sulfur, or novel chemical shapes such as plastics.

Planetary boundaries have been identified by Johan Rockström and his colleagues to define a safe operating space for humanity. However, three of the nine planetary systems identified have been exceeded: Climate change, biodiversity loss and biogeochemical flow. The rough estimates for the remaining six—ocean acidification, land use, freshwater, ozone depletion, atmospheric aerosols, chemical pollution—seem to be 'safe.' Of course, they interact in complex ways that are not well understood.

Furthermore, there may be other boundaries, many more than nine; forest cover is important for a forest planet. Since some of these thought experiments consider and use the net primary productivity of ecosystems, global productivity is a crucial indicator. The health of human populations is an important measurement, taken with unsustainable population size and levels of inequity and other factors. The atmosphere and hydrosphere are constantly forcing raw numbers to fluctuate. There are other boundaries in the ocean, for instance, levels of keystone species. Human pollution is a different kind of boundary from those of natural systems.

At some point the planet may not produce enough oxygen for the oxygen cycle; other cycles will slow or start to fall apart. Terrestrial plants may start to die off, along with their insect and animal obligates. The atmosphere may produce less ozone, which would expose life to more deadly solar rays. Many kinds of life would seek the shade of forests or caves. Others would simply mutate or die or both. The content, workings and shapes of ecosystems would become unpredictable. Things would likely be much more difficult for human beings.

We could avoid a situation of approaching absolute minimal limits, if we invented a steady-state economy, which could be developed indefinitely to be more complex and satisfactory. We can foresee a dreary outcome to 'business as usual,' even if it makes economic sense to drive quickly straight to the cliff, so to speak. We could admit that insane growth is a failure and try to reach a steady state by using our big brains to plan and design an alternate future, where there is enough to support a sophisticated global civilization within a wild planet.

The Inner Limits of Humanity

As members of cultures and physical beings, humans have very real limits. Physical limitations are well-known. We cannot run or jump as fast or far as we want, for instance. We cannot lift too great a weight. We seem to be allocated only a finite number of heartbeats, according to Isaac Asimov, in a single lifetime.

There seem to be mental limits, also. We are unable to remember everything that impacts us. We are unable to give attention to everything at once. The upper limit for measurable relations seems to be about seven (plus or minus two). George Miller found that seven was a magic number in human psychology; it represented the maximum number of items that a subject could reliably remember, as well as other variables. Possibly it applies to the number of subjects having an intelligent conversation as well. In general there may be limits of receiving or processing information.

Some of our problems are the outward manifestations of inner causes, as Ervin Laszlo argues. Our perception and thought structures have limits. Science is logically limited (with a predicate logic), committed to the simple operation of cause and effect and the idea that things cause other things in a linear way. Analytic science has already reached its limits. Data and information developed by hard studies have undercut the paradigms that guided their investigation. The compartmentalization of scientific fields has exposed the complex connections.

There seem to be limits to our personal space and levels of tolerance to human intensification, also. Urban intensification leads to the question: Is there a limit to human numbers? Perhaps there is a limit in terms of space, but is there a psychological limit? People in cities seem to do well with high-contact, high-proximity living. What happens when people are crowded or feel crowded? Physical complaints, emotional complaints, sexual dysfunction, or feelings of fear, seem to be expressed often. There may be limits of crowding. Are there social limits, in terms of the number of people one can tolerate? We may have a requirement for personal space, home space, and wild space.

Psychological limits may be the basis for some of the great failures of human life. The first is the 'failure of perception.' We cannot see slow change or anticipate it. No one really sees the incredible interdependence of humanity and nature, of diversity and stability. We do not seem to be able to see others as feeling human beings. Then, there is the 'failure of imagination,' that limits our understanding and visions of future. We can explore planets and modify genes, but cannot seem to offer functional education or meaningful jobs, dignity in retirement, or goals for living. We insist on individual rights and freedoms, but neglect the whole framework for individual success. We cannot imagine the importance of difference or challenge. Our international system is going to produce a boring uniformity and a painful collapse.

Humans can even create virtual worlds by limiting or expanding what could be received, for instance, if a being could see in the x-ray part of the spectrum. Yet, human imagination is limited, as is human knowledge. Many organisms exist of which we know nothing. Their worlds have little meaning in a human world. E. Hall's space bubble or Lewin's personality field extrapolate the animal or plant *umwelt* to humans. The 'failure of feeling' keeps values and morality as local effects. We do not extend respect or love to distant others in foreign cultures. Our personal values and beliefs do not let us. Finally, the 'failure of nerve' (or will) dooms us to cleave to the familiar, to ignore other alternatives, to fear change or equality.

Genetic Limits to Humanity

Our genetic make-up predisposes us to some things and pushes us in other directions. It does make limits on our plasticity (for possible limits, see Table below: Genetic Behavior Potentially Relevant to Environmental Problems. After G.T. Gardner and P.C. Stern, 1996). It could promote behaviors that damage the environment, and hence our long-term interests. If behavior is limited by 'stone-age' genes, then some pro-environmental behaviors may be ineffective, while others are effective. For example, we must have some contact with the natural environments where we evolved or suffer some psychological and physical harm; this is suggested by several hospital studies. In general, however, behavior is determined by immediate personal consequences, that is, short-term egoism, regardless of eventual consequences in modern world.

The limited genetic potential of a species limits its success in general. Species limitation ensures the diversity and integrity of the whole. A species that was too successful, perhaps like humanity, might endanger the interactions of many other species.

Rene Dubos	Humans have a genetically based need for the stimuli from a natural environment; absence is harmful
Paul Ehrlich	Humans have genetically-based urges for sex and reproduction that cannot be limited, causing overpopulation
B. F. Skinner	Genetically-based "short-term egoism" leads to the environmental Tragedy of the Commons
Edward Wilson	Egoism is tempered by a genetic tendency to live in groups and to behave altruistically towards kin (extended egoism)

Garrett Hardin	A genetically-based denial causes underestimation of probability and severity of environmental threats
Robert Ornstein	Our old mind does not perceive or respond to gradual environmental deterioration
Jay Forrester	The mind cannot comprehend the complexity of social systems, which act in counter-intuitive ways

Perhaps some of these limits have to do with the structure and history of our brain. The bilateral vertebrate, mammalian brain offers an important advantage; it allows two tasks to be addressed simultaneously. Recent brain research (P.F. MacNeilage, 2009) indicates that the left hemisphere specializes in top down control and self-motivated behavior, usually well-established patterns, while the left allows bottom-up control, which is environmentally motivated. The divide allowed specialized behaviors, such as language and tool making, to evolve. Surprisingly, the left hemisphere responds to global patterns to detect and respond to unexpected stimuli. And, the right hemisphere integrates local details. Apparently, immediate global perceptions of change and threats do not translate into true global phenomena, that is, those affecting the planet with slow, invisible, large-scale changes. Perhaps the brain could be trained to respond to those, after recognizing and being affected by planetary change.

Cultural Limits to Humanity

The very circumstance that makes each world-image (cosmology) of a human culture unique—being in a unique place—ensures that it has limits. The image can be tuned to the limits of the local ecology, within their knowledge of interactions. Some cultures ignore the long-range ecological consequences of drainage, irrigation or overexploitation, and these cultures may decline and be extinguished. But many archaic cosmologies are a form of fitness and limitation. Like the Tukano people, most try for adaptation before domination, according to Gerardo Reichel-Dolmatoff. Many of their rituals limit the hunting of fish and game. The entire cosmology of a culture is concerned with some form of adaptation. Particularly in agricultural societies, cosmologies are gauged closely to seasons.

Cosmologies make the world manageable by limiting it. Furthermore, Jeremy Rifkin states that cosmologies are a way of hiding the unimaginable: Extensive voids, confusing gaps, or sheer size. A cosmology can relieve apprehension. They make the world manageable by limiting it. They make the world comfortable and small. Rifkin claims that humanity inflates its daily activity into its image of the universe.

By being limited and contradictory, all cosmologies cause

destruction and waste; all sometimes produce the opposite of the good intended. Archaic and modern, occidental and oriental, world views are complementary, but not complete. Cosmologies can influence a culture to accept or ignore ecological limits.

Once powerful cosmological ideas are adopted, they can influence many cultures over centuries. The principle of plenitude, restated in Christian terms, says that an intelligible creator gave an earth of unlimited bounty to humanity for their use. This principle seemed to be confirmed in the Renaissance with the discovery of the richness of heaven, of microscopic life, and of unexplored continents. Many modern political ideologies and economics have been shaped by the principle of endless wealth. The recognition of limits invalidated it.

The frontier of myth no longer exists. A new myth has to be stronger, the myth of participating in organic beauty, where development, not growth, is without limit. Once doubt is sown—let another recession do that—then the cognitive dissonance from poverty proclaiming it is wealth will transform the old myth. The universe is a frontier, the mind is a frontier, but they are based on a whole and healthy planet, of which humanity is a special part among special parts.

Our cosmologies influence how we respect or exceed such carrying capacities. If for instance a global culture had a cosmological image of the earth as a desert planet, the carrying capacity for humans would probably be reduced to only 100 million people. If a global culture saw itself as completely technological, it would consider that technology could extend the capacity to 10 billion through conversion and substitution. Many modern cosmologies, modified by advances in technology, pretend that the limits can be exceeded. The cosmology of biotechnology is still economic in a primitive sense. Only the myths have changed slightly, to include greater manipulation. It is still concerned with utility, growth and efficiency, as short-term goals. The problem with efficiency is that it is defined within such narrow limits. True efficiency means continuity over long periods of time, e.g., natural processes. Long-term exchanges in nature are not efficient in the industrial sense. A cultural value of slowness would aid survival. Recognizing the limits of the human imagination and technological power would allow us to plan a global economy within the limits of the ocean and land. If we can create an ecological, spiritual image to guide us.

Can We Exceed the Limits?

There are many ways to deal with limits. One obvious way is to ignore them. Ignoring limits, until it is too late to adjust to them, usually results in cultural collapses, as happened to many cultures. Recognition of limits, however, does not mean that we are forced to live as bacteria. Limits allow a considerable amount of freedom within their numbers.

Another way to deal with limits is to try to expand them. Eric Jantsch states that limits to growth can be overcome by the evolution of dynamic structure; they appear as widened limits. Expansion has happened before as the result of evolution. To some extent, technology can expand limits.

Climate change may be a case where we cannot ignore, adjust or expand the threshold. What is the solution for climate change? Restore forests and grasslands, use alternative energy. Should we try a high-tech solution? A massive program, like the atom bomb, only for carbon fixation and alternative energy technologies, might work. Global warming could lead to reversal of ocean currents and a catastrophic sea-level rise.

Limits to growth, even talk of limits, is regarded as a defeatism, as pessimistic, and as a brake to human growth. Unseen limits have as real effects as seen limits. Denying limits does not make them go away. Furthermore, as William Catton has pointed out, there is a difference between raising the limits of carrying capacity and simply permitting greater overshoot of the limits, with the threat of a greater and more catastrophic collapse.

Limits define every system. Limits define locality, local spaces, and local systems, from the global. Limits are not only important for life, but also are implicated in diversity and maturity. But, even these limits are based on the physical.

Although physical and chemical limits are real limits, they are aggravated or increased by psychological limits, which are aggravated by sociocultural limits, which are further increased by political limits. Each 'higher' more complex level makes acceptance of other levels more difficult. The attempt to exceed limits becomes less efficient. One of the most important psychological limits is our ability to process data and to draw conclusions from it. We are so ignorant of the complexities of ecosystems that it is suicidal to pretend to 'maximize' their use for resources. A free market civilization has to be limited by conservative calculations of ecological balance. It is almost impossible to estimate the economic value of this natural balance. Some limits cannot be exceeded and have to be applied to wilderness or to human populations.

Or, we can respect limits and work within them. This is an ethical way. We have been concerned only with the present; posterity has been ignored. Human duration is too short. The systems and processes that we deal with are intergenerational, which is an additional challenge.

Our regard for the challenge of intergenerational economic limits is irrational. Future values are discounted at low rates, so we can ignore them. Planting a tree that takes 30-200 years to mature is calculated to be uneconomical (the giant sequoia takes almost 200 years to flower, for instance). Who benefits from it? If our ancestors had planted trees for our benefit, it would be easier to justify continuation. But identification through time is related to that of place. People not living in place have no vested interests.

The land ethic Leopold described was a sense of ecological community between humanity and other species. When we see land as community to which we belong, we will use it with love and respect. Such an ethic would change the human role from master of earth to plain member of it. Predators are important members of the community; no special interest has the right to exterminate them to benefit itself. This attitude is important for habitat protection. Aldo Leopold describes the extension of ethics as a process in ecological evolution. Its sequences had to be described in ecological as well as in philosophical terms. An ethic, ecologically, has to be a limitation on freedom of action. An ethic, philosophically, is a differentiation of social from anti-social conduct. These are two different perspectives of one ethic, which has its origin in the tendency of interdependent individuals or groups to evolve modes of cooperation.

Humanity is part of nature, as valuable and unique as cranes or lousewarts, but not more valuable or unique. This attitude would entail using what is necessary, exploiting some ecosystems completely, changing a place to fit human aspirations, and killing plants and animals for sustenance. All animals and plants alter their environments to fit to some extent. But it would also mean limiting humanity and its technological effects, limiting human use to local impacts, and letting other beings live without interference. It is not necessary to dominate or terraform the earth completely. Humanity could contain itself to five percent of the earth's surface and ecosystems and only visit or ignore the remainder. This ideal requires change, even different ways of treating nature.

We could respect limits by being wise, or acting 'as if' we were wise. Jonas Salk defined wisdom as the art of disciplined use of imagination in respect to alternatives, exercised at the right time and in the right measure. Judgment is required as to what is right, and judgment may be an innate art. It is a new kind of fitness, supplanting the biological kind of evolution. Humans have made radically different conditions that they must now accommodate. If the mind is exposed to economy of nature, as revealed through living systems, humans may recognize the necessity of balancing values. Total win-lose conflicts are unwise. Value systems concerned with dynamic equilibrium, aesthetics, complementarity, reciprocity, interdependence, reconciliation, and intuition–*this is the language that ecology speaks.*

The New Government Office of Jester

Our representatives, mayors, governors, newsreaders, 'talkshowhosts,' and doomsayers appear so serious. They are arrogant, glum, stiff, vituperative, and unwilling to consider any other views, information, facts, ideas, or opinions. There must be some way to lighten their burden of defensive layers. Laughter perhaps.

Kings formerly had jesters to entertain them (called fools in the middle ages), and they appeared in many European, African and Asian countries. Kings were often insulated from reality and truth, much like many well-off politicians today ("Gosh, I'm employed; let those welfare spongers get jobs and buy food"). The jester was the only one who could anchor the ship of state in charted waters. Jesters could sometimes deliver to kings the bad news that might cost others their heads, the Queen's dalliances or the loss of a battle. Once, centuries ago, even British households hired licensed jesters to amuse as well as to criticize masters and guests. This was accepted, in a limited way, because they all, kings, royalty and pretenders, acted as if the jester was moronic and mad, and he couldn't help it. The jester also had the lowest status in families or courts and was exempt from the normal rules and social expectations. Jesters had to be amusing, but were expected to be honest in observing and commenting on the behavior of their betters. They had the power to mock and revile the most prominent, until they were removed or killed, for being too accurate or insistent.

We need to institute that office again, but maybe without the possible death penalty. In the comedy of politics, the jesters would present common sense and truth. A new Office of Jester is crucial to the continuity of modern government, only now it is the little elected kings who are moronic and mad. Politicians have learned that nothing is too dumb or extreme, regardless of how ridiculous or socially destructive it is. As long as a few of their constituents, probably family members under 18, love their idiocy, cute greed and smiling shrewdness, they continue it. Since politicians likely do not see *The Daily Show* or read the comics, it might help if they had a personal Fool to remind them of their intellectual and moral errors, not only to protect us from losing our health and paltry wealth, but to protect them from pangs of conscience or moral meltdown. But, are they smart enough, humble enough, to act on such a reminder? The Supreme Court certainly needs a Court Jester. Where is the Irony Board to address the wrinkles, from the crooked crush of greed and stupidity, in the fabric of social cohesion? This government by 'make-believe' is creating a thin cloud too far from the ground—dare one say 'Cloudcuckooland'? With apologies to Aristophanes.

It is urgent that we create a formal position of Jester to attend every high political position and ground it. Let's not just fool around. It might cost something, but it would employ many unemployed comedians or funny philosophers. They could lightly amuse the 'Decisioners' and then deliver the really bad news: Shrinking economies, social upheaval over massive inequities, dangerous armaments not for warfare, and runaway environmental destruction and looming catastrophes. Basically the message is: *Stop pretending all is right and business as usual can save the day—It's an emergency!* We all need to act. Won't someone considered important and alert, or entertaining and honest, please tell our political leaders that *the planet is burning!?*

Is Humanity a Cancer in the Planet?

How can we look at humanity? First we were wolves, catching animals and eating them. For thirty thousand years. Then we were cows, standing in place, eating grasses and pooping, for ten thousand years. Now we are termites, swarming over everything in furious dances of labor and status, for the past five thousand years. We still poop as we walk, but the energy slaves flush it downstream.

Are we urban by nature? Have we changed from hunting to urbaning? Of course, we may be preadapted for cities. We are clever social animals. We prefer edge habitat so as to move between other habitats. Urbanization is a characteristic environment of an edge species, modified by us to suit our desires and fit our behavior. Civilization produces edges that people like. Wilderness is fragmented into more edges, in islands and patches. So, there is no more deepness, no more interior to wilderness. What are other edge species? Raccoons, coyotes, crows, and rats.

We are still foragers and predators at heart. What are the needs in human nature that a city answers? Perhaps the city is what wilderness was—a place of passage, a place to be brave and test yourself. Humans have always worked to abstract their specialness. The city seems to be another myth of separateness and independence from nature. The city is the laboratory of human creativity, kept apart from the mother of nature.

By their activities, human beings change the places they live. Much of the change is easily incorporated in the cycles of renewability of the ecosystems. However, humans often change the directions of such systems by simplifying or degrading them. Here, humans act as agents of interference.

Humans have had a great impact on nature, and should be considered as a force of nature and history. Humans are special agents that act like parasites, drawing nourishment and weakening the host. States acted like macroparasites, according to William McNeill, but became less violent or unpredictable over time, as they adjusted to their host populations.

As crowding increases competition, there is more pressure on the remaining reserves. The system itself parasitizes humanity and nature. Humanity becomes an autoparasite, a new pseudo-species. Technology enlarges the number of niches for us; tools fit humans to different habitats, displacing other species. We steal from animals and plants, from the earth, and from our own descendants. Hobbes foresaw this war of each against all. The systematic destruction of human beings and animals is not an isolated peculiarity. A fat parasite often kills it host and then dies itself. Perhaps, humanity is a systems agent that encourages only positive feedback.

Perhaps human expansion is like a cancer. Alan Gregg compared the world to a living organism and the explosion in human numbers to the proliferation of cancer cells. He sketched other parallels between cancer in humans and humans' cancer-like impact on the world. Cancer cells proliferate rapidly and uncontrollably in the body; humans continue to proliferate rapidly and uncontrollably in the world. Crowded cancer cells harden into tumors; humans crowd into hardened cities. Cancer cells infiltrate and destroy adjacent normal tissues; urban sprawl devours normal open land. Malignant tumors shed cells that migrate to distant parts of the body and set up secondary tumors; humans have colonized just about every habitable part of the globe. Cancer cells lose their natural appearance and distinctive functions; humans homogenize diverse natural ecosystems into artificial monocultures. Malignant tumors excrete enzymes and other chemicals that adversely affect remote parts of the body; humans' motor vehicles, power plants, factories, and farms emit toxins that pollute environments far from the point of origin.

It is not in a tumor's self-interest to steal nutrients to the point where the host starves to death, for this kills the tumor as well. Yet tumors commonly continue growing while the victim wastes away. A malignant tumor usually goes undetected until the number of cells in it has doubled at least thirty times from a single cell. The number of humans on Earth has already doubled thirty two times, reaching that mark in 1978 when world population passed 4.3 billion. It is over seven billion now. After thirty-seven to forty doublings, at which point a tumor weighs about one kilogram, the condition is usually fatal—that would be the population equivalent of 5.4 billion people. We have exceeded that; the question is, is it fatal—large complex systems may take a long time to collapse.

Every species exploits its environment to the extent that it can, with no regard to consequences. Usually, each species is checked by another, because there are many competing for the same food, and a loose equilibrium is maintained. By distorting the equilibrium, we have destroyed whole species and favored many others.. It is not that our actions are different from other species; it is that they are excessive, rapid, compounded, and large-scale. We are a force to be contained, voluntarily or not.

We're All Shrinking

Many animals and plants are getting smaller. This seems to be a long-term trend associated with the interglacial warming. Current bison and wolves, for instance, are much smaller than their predecessors. Polar bears are getting smaller. They may be evolving to tolerate warmer conditions and smaller ice flows, although it may be literally too little too late. Polar bear skulls from the latter half of the twentieth century were 2 to 9 percent smaller than those from the early part of the century. The changes could be linked to an increase in pollution and the reduction in sea ice. Because the ice is melting, the bears have to use more energy to hunt. This increased effort needed to find food likely limits the animals' growth. There are other possible causes: It could be related to a reduction in genetic diversity; hunting over the last century may have depleted the gene pool, leaving polar bears to suffer the effects of inbreeding; it might be that the shape changes are related to pollutants, especially halogen compounds such as fluorine, chlorine, bromine or iodine, that have built up in the Arctic and in the polar bears' bodies. Many substances are still being used in solvents, pesticides, refrigerants, adhesives, and coatings (because we don't think or ban them). Polar bears just seem to attract pollutants.

Tortoises, common toads, and iguanas have all started to reduce in size over the past 200 years. Global warming seems to have speeded up the shrinking. Of 85 species studied by scientists, almost half have been shrinking slowly, e.g., Red deer in Europe and sheep in Scotland.

We would expect plants to be growing bigger as more CO_2 enters the atmosphere. But they, too, are shrinking, possibly due to droughts from climate change.

It's been shown that rising CO_2 levels in the atmosphere make ocean water more acidic, which causes plankton, corals, and mollusks to decrease in number. This lessens the amount of food available to species, which filters all the way up to the top of the food chain. Those reduced food supplies are likely to mean that animals at the apex—including humans—will change.

There are benefits to shrinking. Better fitness to food supply; each requires less food to be healthy. Larger populations are possible in place.

There are hazards as well. Amphibians could dry up faster in higher temperatures in one season. Other species, even mammals will have fewer offspring, and be more vulnerable to disease. As Polar bears weight has fallen, their health has suffered, impairing their ability to reproduce and have cubs that survive. And some scientists report that, in their desperation, polar bears are turning on each other. They deliberately hunt other bears. Also, as certain types of flora and fauna diminish or grow smaller in size, we'll all have to work harder to procure food.

During previous warm eras, e.g., 55 million years ago, invertebrates like beetles, bees, and ants became 50 to 75 percent smaller. Other species ranging from single-celled diatoms to woodrats have also been shown to decrease in size. Is it natural, then? Regardless, many cold species will disappear. If a large number of species gets smaller at the same time, our new, tinier world of plants and animals will be different.

Human culture can change conditions faster than biology can respond. Shrinking in size may become true of us as well. That change in size could be beneficial for city living, allowing higher densities. One should ask if human goals and horizons are also getting smaller, as we focus on entertainment and immediate rewards. Will we even notice the great shrinking?

Think Long Term

Trading for short-term security leads to fewer survival options and less flexibility. In fact, it leads to a trap. One of our problems is our inability to think over long, long time lines. So, we adapt to short-term catastrophes and challenges, but become more vulnerable to long-term ones. Politicians and corporations tend to be myopic, limited to 2 or 4 year horizons. Because everyone seems focused on this horizon, it can lead to a collective trap.

We do not see the full range of possibilities, nor do we see the impacts of our immediate activities on the system. Herbert Simon judges we are not omniscient rational optimizers, but rather blundering satisficers. We satisfy our needs well enough before moving on to next challenge. Thus, we tend to make a choice we can live with now, rather than work for some long-term optimum. We do not correctly interpret imperfect information. With no long-term thinking, wars become more likely and environmental damage becomes more likely.

The kind of enlightened self-interest that leads people to cooperate requires an ability to think in the long-term and to visualize long-term forms of traps, such as addiction, policy resistance, arms races, drifts to low performance, and the tragedy of commons. Perhaps instances of long-term planning were influenced by special challenges or relatively long periods of cultural stability. Swedish long-term planning may have been affected by long winters; planning in Israel may be supported by the long-term goal of survival. Several longer-term plans in China may be the result of its cultural stability. Some nations, such as Costa Rica or Switzerland, have created long-term plans for conservation or village growth. It may not necessary to be rational optimizers; we can be blundering satisficers and still consider the full range of impacts and possibilities. We can live with uncertainty and misperceive risk and still survive in a satisfactory way, as long as we understand and are attentive to long-term trends.

If we think of long-term trends in terms of growth, as cities have, or long-term trends of collapse, the trends do not seem sustainable for long. It may be that we can use current ideas of development, rather than growth, as a switch to maturity for human progress. In other words, our species may be becoming mature and consciously developing instead of blindly growing. Development makes long-term thinking easier.

Nature tends to select harmony and cooperative behaviors in the long-term, by filtering out less adaptive behaviors and strategies. If we understand the dynamic interactions between human societies and their environments, from the perspective of long-term patterns and processes, then we see that nature works with pulsed exchanges, in the short and long-term, with regular and irregular patterns. Pulses have to be intrinsic to any design, that is, the design has to accept chaotic energy patterns. The scale has to fit. Maybe strict conservation is the best path. Long-term solutions to water and food shortages may require conservation.

If we were capable of long-term thinking, as well as of denial and satisfactory development, we could make more effective, anticipatory designs. Design may take a long time—longer than human lifetimes. So, designing and planning has to be a long-term thing. We could start with a long-term ecological perspective, assuming traditional strengths, but not planning the future in detail. Or could we just start the framework to shape the trends.

By emphasizing quality, ecological design contributes to long-term products and patterns. Design can make models of quality environments. By embracing ideas from ecology, design participates in a movement of consciousness, concerned with equality, diversity, and health, as well as with humane methods, and a holopoetic cosmology—and is affected by them simultaneously. Thinking in the long-term is a form of play. Play can be an experimental dialogue with the environment, by repeating or recombining sequences of behavior outside their primary context. And, *it can be fun.*

Figure. River cleanup in Aprilci. Then the Frisbees came out.

Select Bibliography

Aberle, Doug, Editor. 1994. The Future by Design: The Practice of Ecological Planning. Gabriola Island, BC, CA: New Society Publishers.

Alexander, Christopher, et al. 1977. A Pattern Language. New York: Oxford University Press.

Appadurai, Arjun. 1996. Modernity at Large. Cultural Dimensions of Globalization, Minneapolis/London, University of Minnesota Press

Arbib, Michael. 1972. The Metaphorical Brain. New York, John Wiley and Sons.

Avise, John C. 1994. "Conservation genetics," in Molecular Markers, Natural History and Evolution. New York: Chapman & Hall.

Axtmann, Roland (1995), Kulturelle Globalisierung, kollektive Identität und demokratischer Nationalstaat, Leviathan 23 (1), 87-101.

Bachelard, Gaston. 1971. On Poetic Imagination and Reverie. trans. C. Gaudin. Indianapolis: Bobbs-Merrill Co. Inc.

Bain, M. B. and J. T. Finn. 1988. Streamflow regulation and fish community structure. Ecology 69(2):382-392.

Baker, Richard St. Barbe. 1980. "A man of the trees: Ed Goldsmith interviewing Richard St. Barbe Baker." Coevolution Quarterly 25: 66-70.

Barfield, Owen. 1964. Poetic Diction: A Study in Meaning. New York: McGraw-Hill.

Barton, Hugh, Editor. 2002. Sustainable Communities: The Potential for Eco-Neighborhoods. London: Earthscan.

Bateson, Gregory. 1987. Steps to an Ecology of Mind. Northvale, NJ: Jason Aronson Inc.

Beatley, Timothy. 2004. Native to Nowhere. Washington: Island Press.

Bell, J. S. 1964. On the Einstein Podolsky Rosen Paradox. Physics 1:195-200.

Bellah, Robert N. et al. 1991. The Good Society. New York: Alfred A. Knopf.

Bergstraesser, A. 1962. Goethe's Image of Man and Society. Freiburg: Herder.

Berry T., The Dream of the Earth, San Francisco, CA, Sierra Books, 1988

Berry W., The Unsettling of America: Culture and Agriculture, New York, Avon, 1978.

Bertalanffy, Ludwig von. 1952. Problems of Life: An Evaluation of Modern Biological Thought. New York: John Wiley and Sons.

Bertalanffy, Ludwig von. 1975. Perspectives on General Systems Theory. New York: G. Braziller.

Bhabha, Homi K. (1996), Culture's In-Between, in Stuart Hall/Paul du Gay (Ed.) Cultural Identity, London, Sage, 53-60.

Birch, Charles, and Cobb, Jr., John B. 1981. The Liberation of Life. Cambridge: Cambridge University Press.

Birkeland, Janis. 2004. Design for Sustainability: Sourcebook of Integrated Ecological Solutions. London: Earthscan.

Black, Max. 1962. Metaphor In: Models and metaphors. M. Black, Ed. Ithaca: Cornell University Press.

Bohm, David. 1980. Wholeness and the Implicate Order. London: Routledge & Kegan Paul.

Boltzmann, Ludwig. 1974. The Second Law of Thermodynamics. Populare Schriften.

Borgstrom, George. 1965. The Hungry Planet. New York: Macmillan Co.

Boulding, Kenneth E. 1956. The Image: Knowledge in Life and Society. Ann Arbor: University of Michigan Press.

Brand, Stewart. 2009. Whole Earth Discipline. New York: Viking.

Brown, George Spencer. 1969. The Laws of Form. London, Allen and Unwin.

Brown, Lester. 1979. "Crossing the threshold?—Pressures on earth's biological systems." Environment 21(8):12-37.

_____. 2003. Plan B. Rescuing a Planet Under Stress and a Civilization in Trouble. New York: W.W. Norton.

Buchanan, Richard. 1992. Wicked problems in design thinking. Design issues. 8:5-21.

Bunge, Mario. 1973. Method, Model and Matter. Boston: Synthese Library.

Burlingh, P. et al. 1975. Computation of the Absolute Maximum Food Production of the World. Wageningen, Netherlands: Agriculture University.

Butler, Tom, Ed. 2002. Wild Earth. Minneapolis: Milkweed.

Carlson, Christine and Dennis Canty. 1986. A Path for the Palouse. Seattle: National Park Service.

Carson, Rachel. 1962. Silent Spring. Boston: Houghton Mifflin.

Catton, William R. 1982. Overshoot: The Ecological Basis of Revolutionary Change. Urbana: University of Illinois Press.

Cheng, T.C. 1970. Symbiosis. New York: Pegasus.

Chew, Geoffrey. 1970. Hadron bootstrap. Physics Today Oct., 1970: 23.

Chew, Sing C. 2006. World Ecological Degradation. New York: Altamira Press.

Cherfas J., and M. & J. Fanton, The Seed Saver's Handbook, Chichester, UK, Grover books, 1996. See also International Pant Genetic Resources institute www.ipgri.cgiar.org

Chernushenko, David. 2008, July 28. A model for real community energy self-sufficiency. Times Colonist, (op ed section). See his project at www.livinglightly.ca/film.

Clark, Colin. 1958. "World Population." Nature 181:1235-1236.

_____. 1967. Population Growth and Land Use. London: Macmillan.

Cobb, John B., Jr. 1965. A Christian Natural Theology. Philadelphia: Westminster.

_____. 1971. Is It Too Late? A Theology of Ecology. New York: Glencoe.

_____. 1992. Sustainability. Eugene: Wipf & Stock.

_____. 1994. For the Common Good. Boston: Beacon Press.

_____. 2010. Spiritual Bankruptcy: A Spiritual Call to Action. New York: Abington Press.

Conservation International and Agrupación Sierra Madre. "Wilderness: Earth's Last Wild Places."

Cooper J., Leifert C, and Niggily U., Eds. 'Handbook of Organic Food Safety and Quality', Cambridge, UK 2007.

Costanza, Robert, Ed. 1991. Ecological Economics: The Science and management of Sustainability. New York: Columbia University Press.

Costanza, Robert, Bryan G. Norton, and Benjamin D. Haskell, Eds. 1992. Ecosystem Health: New Goals for Environmental management. Washington: Island Press.

Costanza, Robert, L.J. Graumlich, and W. Steffen, Eds. 2007. Sustainability or Collapse? Cambridge: MIT Press.

Cummins, K. W. 1979. "The natural stream ecosystem," in J. V. Ward and J. A. Stanford, Eds. The Ecology of Regulated Streams. New York: Plenum.

Daly, Herman. E. 1977. Stead-State Economics. San Francisco: Freeman.

_____. 1968. Economics as a life science. Journal of Political Economy 76:392- 401.

Daly, H.E. and Cobb, J. B., Jr. 1989. For the Common Good: Redirecting the Economy Toward Community, the Environment and a Sustainable Future. Boston: Beacon Press.

Dansereau, Pierre. 1957. Biogeography: An Ecological perspective. New York: Ronald Press.

Darwin, Charles. 1962. The Origin of the Species by Means of Natural Selection or the Preservation of Favoured Races in the Struggle for Life. New York: Collier.

Dasmann, Raymond. 1972. Environmental Conservation. New York: Wiley.

Daubenmire, Rexford. 1942. An ecological study of the vegetation of southeastern Washington and adjacent Idaho. Ecol. Mono. 12:53-79.

DeWit, K. 1967. Incomplete reference.

Diamond, J. M. 1975. The island dilemma: Lessons of modern biogeographic studies for the design of natural preserves. Biol. Conserv. 7:129-146.

_____. 1997. Guns, Germs, and Steel: The Fate of Human Societies. New York: W. W. Norton.

Diamond J. 2005. Collapse: How Societies Chose to Fail or Succeed. New York, Penguin.

Dobben, W.H. van, and R.H. Lowe-McConnell, Eds. 1975. Unifying Concepts of Ecology. Report of the Plenary Sessions of the First International Congress of Ecology. The Hague.

Dobson, A. P. et al. 1991. "Conservation biology: The ecology and genetics of endangered species," in Genes in Ecology, R. J. Berry et al., Eds. London: Blackwell Scientific.

Domingo J.L. Toxicity Studies of Genetically Modified Plants: A Review of Published Literature, Critical Reviews in Food Science and Nutrition 47:721-733, 2007.

Doolittle, W. F. 1981. Is nature really motherly? Coevolution Quarterly 29:58-62.

Doxiades, C. A. 1975. Building Entopia. New York: Norton.

_____. 1977. Ecology and Ekistics. Gerald Dix, Ed. London: Elek Books.

Drengson, Alan. 1980. Shifting Paradigms: From Technocrat to Planetary Person. Environmental Ethics, 3, 221-240. Revised in Drengson and Inoue, 1995.

_____. 1989. Beyond Environmental Crisis: From Technocrat to Planetary Person. New York: Peter Lang.

Drengson, Alan et al. 1994. "The ecoforester's way: An oath of ecological responsibility," International Journal of Ecoforestry 10(1):48.

Drengson, Alan. (1995). The Practice of Technology. Albany NY: SUNY Press.

Drengson, Alan and Yuichi Inoue (Eds). 1995. The Deep Ecology Movement: An Introductory Anthology. Berkeley, CA: North Atlantic Books.

Drengson, Alan and Duncan Taylor. 1997. Ecoforestry: The Art and Science of Sustainable Forest Use. Gabriola Island, BC: New Society Pub.

Drucker, Peter. 1990. The New Realities. New York: Dutton.

_____. 1993. The Post-Capitalist Society. New York: HarperCollins.

_____. 1995. Managing in a Time of Great Change. New York: Dutton.

Dubos, Rene. 1965. Man Adapting. New Haven: Yale University Press.

_____. 1976. Symbiosis between the earth and humankind. Science 193:459-462.

_____. 1980. The Wooing of Earth. New York: Charles Scribner's Sons.

Dyson, Freeman. 1979. Disturbing the Universe. New York: Harper & Row.

Easterbrook, D. J. and D. A Rahm. 1970. Landforms of Washington: The Geological Environment. Bellingham: Union Printing Co.

Eckholm, Eric P. 1976. Losing Ground: Environmental Stress and World Food Prospects. New York: W. W. Norton.

Edie, James. 1972. New Essays in Phenomenology. New York: Quadrangle.

Ehrlich, P. and A. Ehrlich. 1972. Population, Resources, Environment. San Francisco: Freeman.

Ehrlich, P. 1981. An ecologist standing up among seated social scientists. Coevolution Quarterly 31:24-35.

Eibl-Eibesfelt, I. 1970. Ethology: The Biology of Behavior. New York: Holt, Rinehart and Winston.

Eigen, Manfred, and P. Schuster. 1979. The Hypercycle: A Principle of Natural Self-Organization. Berlin: Springer-Verglag.

_____. 1981. Laws of the Game. New York: Alfred A. Knopf.

Einstein, Albert and Leopold Infeld. 1966. The Evolution of Physics. New York: Simon and Schuster.

Eisenberg, Evan. 1998. The Ecology of Eden. New York: Vintage.

Eliot, T.S. 1948. Notes towards a Definition of Culture. London: Faber & Faber.

Ellen, Roy and Katsuyoshi Fukui, Eds. 1996. Redefining Nature. Oxford: Berg.

Elton, Charles. 1966. Animal Ecology. New York: October House.

Emmons, H. 2006. The Chemistry of Joy. New York: Simon and Schuster.

Eschenbach, Ted G. and G. A. Geistauts. 1986. Alaska's Future: Commentary on a Delphi Perspective. Alaska Pacific University Press.

Evernden, N. 1981. Out of Place. unpublished manuscript.

Ewald, P.W. 1993. The Evolution of Virulence. Scientific American, April, p 86-93.

Eyre, Samuel. 1978. The Real Wealth of Nations. London: E. Arnold.

Fiedler, Peggy L. and Subodh K, Jain, Eds. 1992. Conservation Biology. New York: Chapman & Hall.

Flannery, Tim. 2002. The Eternal Frontier. New York: Grove Press.

_____. 2010. Here on Earth. New York: Atlantic Monthly Press.

Forestry Commission, 1994. Forest Landscape Design. London: HMSO.

Foreman, Dave. 1985. Ecodefense. San Francisco: Earth First! Books.

_____. 2004. Rewilding North America. Washington: Island Press.

Formann, R. T. T. and M. Godron. 1986. Landscape Ecology. New York: John Wiley.

Formann, Richard T. T. 1997. Land Mosaics. Cambridge: Cambridge University.

Fox, Michael W. 1974. Concepts in Ethology: Animal and Human Behavior. Mineapolis: U. of Minnesota Press.

_____. 1976. Between Animal and Man. New York: Coward, McCann and Geoghehan Inc.

Fox, M. W. 1980. One Earth, One Mind. New York: Coward, McCann and Geoghehan Inc.

———————. 1978. Personal Communication.

———————. 1980. One Earth, One Mind. Coward, McCann and Geoghehan Inc., NY, pp. 174-234.

———————. 1980 Returning to Eden: Animal Rights and Human Responsibility. Viking Press, NY, pp. 19-141.

———————. 1984. Farm Animals: Husbandry, Behavior and Veterinary Practice' Baltimore MD.

———————. 1988. Agricide: The Hidden Farm and Food Crisis that Affects us All. New York: Schocken books.

———————. 1997. Eating With Conscience: The Bioethics of Food. Troutdale OR: New Sage Press.

———————. 2001. Bringing Life to Ethics: Global Bioethics for a Humane Society. Albany, NY State University of New York Press.

———————. 2004. Killer Foods: What Scientists Do To Make Better Is Not Always Best. Guilford CT: The Lyons Press.

———————. 2011. Healing Animals and the Vision of One Health. Golden Valley: One Health Vision Press.

Frankel, Otto. 1975. Crop Genetic Resources for Today and Tomorrow. O. Frankel and J. Hawkes, Eds. New York: Cambridge Univ. Press.

Frankel, O. H. and M. E. Soule. 1981. Conservation and Evolution. Cambridge: Cambridge University Press.

Friedman, Thomas L. 2005. The World is Flat. New York: Farrar, Straus & Giroux.

Fromm, Erich. 1956. The Art of Loving. New York: Harper.

———————. 1976. To Have or To Be. New York: Bantam Books.

Frome, Michael. 1974. Personal Communication.

Fowles, J. 1979. Seeing Nature Whole. Harper's. 259:49-56.

Fukuoka, Masanobu. 1978. One-Straw Revolution. Emannus, PA: Rodale Press.

———————. 1985. The Natural Way of Farming: The Theory and Practice of Green Philosophy. New York: Japanese Publications Inc.

Fuller, R. Buckminster. 1969. Operating Manual for Spaceship Earth. Lars Muller Pub.

———————. 1970. Utopia or Oblivion. Lars Muller Pub.

———————. 1981. Critical Path. New York: St. Martin's Press.

———————, with E.J. Applewhite. 1982. Synergetics. New York: Macmillan Pub Co.

Gabel, Medard. 1979. HO-PING: Food for Everyone. Garden City, NY: Doubleday.

Georgescue-Rogen, Nicholas. 1971. Entropy and the Economic Process. Cambridge: Harvard University Press.

Gezon, Lisa. 2006. Global Visions, Local Landscapes. New York: Altamira Press.

Goldsmith, Edward et al. 1972. A Blueprint for Survival. Boston: Houghton Mifflin Co.

Goodland, R. 1997. Environmental Sustainability in Agriculture: Diet Matters. Ecological Economics, 23: 189-200.

Golley, Frank B., K. Petrusewicz, and L. Ryszkowski, Eds. 1975. Small Mammals: Their Productivity and Population Dynamics. New York: Cambridge.

Goodman, Paul. 1962. Utopian Essays and Practical Proposals. New York: Vintage.

Gould, S.J. 1977. Ever Since Darwin. New York: W.W. Norton.

Gray, Russell D. 1988. Metaphors and methods. In Mae-Wan Ho and S. W. Fox, Eds., Evolutionary Processes and Metaphors. New York: Wiley.

Greene, Patricia and Dean Apostol. 1994. Design for biodiversity. Landscape Architecture 85 (4):63-65.

Gunderson, Lance H. and C.S. Holling, Eds. 2002. Panarchy: Understanding Transformations in Human and Natural Systems. Washington: Island Press.

Hall, Edward T. 1969. The Hidden Dimension. Garden City, NY: Doubleday & Co.

Hall, Stuart (1992), The Question of Cultural Identity, in Stuart Hall et al., Eds., Modernity and its Futures. London, Polity Press, 273-325

Hampden-Turner, C. 1981. Maps of the Mind. New York: Macmillan.

Hardin, Garrett. 1969. Population, Evolution, and Birth Control. San Francisco: Freeman.

_____. 1977. The Limits of Altruism. Bloomington: Indiana University Press.

_____. 1987. Cultural carrying capacity. Fourth World Wilderness Cong.

_____. 1993. Living within Limits. New York: Oxford University Press.

Harper, J. L. 1977. Population Biology of Plants. New York: Academic Press.

Harris, Larry D. 1984. The Fragmented Forest: Island Biogeography Theory and the Preservation of Biotic Diversity. Chicago: UC Press.

Hart, Richard. 1994. "Monitoring for ecosystem management," International Journal of Ecoforestry 10(2):74-75.

Hartmann, Thom. 2004. The Last Hours of Ancient Sunlight. New York: Three Rivers.

Hatch, L.U. and M.E. Swisher, Eds. 1999. Managed Ecosystems. New York: Oxford.

Hebb, D.O. 1958. Alice in Wonderland or psychology among the biological sciences. In The Biological and Biochemical Bases of Behavior. H. Harlow and C. Woolsley, Eds. Madison: University of Wisconsin Press.

Henberg, M. 1984. "Wilderness as playground." Environmental Ethics 6:253-263.

Henderson, Hazel. 1980. Creating Alternative Futures: The End of Economics. New York: Putnam.

Higgs, Eric. 2003. Nature by Design: People, Natural Process and Ecological restoration. Cambridge: MIT Press.

Ho, Mae-Wan and S. W. Fox. 1988. Processes and metaphors in evolution. In Mae-Wan Ho and S. W. Fox, Eds., Evolutionary Processes and Metaphors. New York: Wiley.

Hoffman, David. 1976. Inuit land use on the barren grounds. Inuit Land Use and Development Project, Vol. 2. Ottawa: Dept. of Indian and Northern Affairs.

Holling, C.S. 1973. "Resilience and stability of ecological systems." Annual Review of Ecology and Systematics. R.F. Johnston et al., editors. 4:1-24.

Homer-Dixon, Thomas. 2006. The Upside of Down. Washington: Island Press.

Hornborg, Alf and Carole Crumley, Eds. 2007. The World System and the Earth System. Walnut Creek: Left Coast Press.

Hornborg, Alf et al. 2007. Rethinking Environmental History. New York: Altamira Press.

Houghton R. A. and D. L. Skole. 1990. "Changes in the global carbon cycle between 1700 and 1895, In The Earth Transformed by Human Action, B. Turner, Ed. Cambridge: Cambridge University Press.

Hu Frank B. and Walter C. Willett. 1998. The Relationship Between Consumption of Animal Products ... and Risk of Chronic Disease: a Critical Review. Report for the World Bank. Cambridge, MA: Harvard School of Public Health.

Hughes, J. Donald. 1983. American Indian Ecology. El Paso: Texas Western Press. Indian use of forests, Pp. 98-100.

Hulet, H.R. 1970. "Optimum world population." Bioscience 20(3):160-161.

Huston, Michael A. 1994. Biological Diversity. Cambridge: Cambridge University Press.

Huxley, Aldous. 1945. The Perennial Philosophy. New York: Harper.

_____. 1956. "Knowledge and Understanding." In Adonis and the Alphabet. London: Chatto and Windus.

_____. 1977. The Human Situation. P. Ferrucci, Ed. New York: Harper & Row.

Illich, Ivan. 1973. Tools for Conviviality. New York: Harper and Row.

Imhoff D., and Baumgartner J.A., Eds. 2006. 'Farming and the Fate of Wildlife Healdsburg, CA: Watershed Media.

Jackson D.L., and Jackson L.L., Eds. 2002. The Farm as Natural Habitat: Reconnecting Food Systems With Ecosystems. Washington DC: Island Press.

Jackson, Wes. 1980. New Roots for Agriculture. San Francisco: Friends of the Earth.

_____. 1994. Becoming Native to this Place. Lexington: University of Kentucky.

_____. 2010. Consulting the Genius of Place. Berkeley: Counterpoint.

Jackson, Wes, W. Berry, and B. Colman. 1984. Meeting the Expectations of the Land: Essays in Sustainable Agriculture. San Francisco: North Point.

Jacobs, Jane. 1969. The Economy of Cities. New York: Random House.

James, A., K. Gaston and A. Balmford. 1999. "Balancing the earth's Accounts." Nature 401:323-324.

Jantsch, E. 1975. Design for Evolution. New York: Braziller.

_____. 1980. The Self-Organizing Universe. New York: Pergamon Press.

Johnson, Steven. 2001. Emergence. New York: Scribner.

Jordan III, W. R., M. E. Gilpin, and J. D. Aber, Eds. 1989. Restoration Ecology: A Synthetic Approach to Ecological Research. Cambridge: Cambridge University Press.

Jouvenel, Bertrand de. 1968. The stewardship of the earth. In: The Fitness of Man's Environment. Smithsonian Annual II. New York: Harper Colophon Books.

Julier, Guy. 2008. The Culture of Design. New York: Sage Publications.

Kaplan, Rachel. 1983. "The role of nature in the urban context," in I. Altman and J. F. Wohlwill, Eds. Behavior and the Natural Environment. New York: Plenum Press.

Kaplan, Robert D. 2000. The Coming Anarchy. New York: Vintage Books.

Kaplan, S. 1973. "Cognitive maps; human needs and the designed environment." In W. F. E. Preiser, Ed., Environmental Design Research. Stroudsberg: Dowden, Hutchinson & Ross.

Kauffman, Stuart A. 1993. Origins of Order: Self-organization and Selection in Evolution. Oxford: Oxford U. Press.

Keeling, C. D. and T. P. Whorg. 1992. Muana Loa: Atmospheric CO_2—modern record. In T. A. Boden et al., Eds., Trends 91: A Compendium of Data on Global Change. Oak Ridge: Oak Ridge National Laboratory.

Kellert, S.R. 1996. The Value of Life: Biological Diversity and Human Society. Washington: Island Press.

Kellert, Stephen R. and E.O. Wilson. 1993. The Biophilia Hypothesis. Washington: Island.

Kelly, Kevin. 1998. New Rules for the New Economy. New York: Viking.

Kelsall, J. P. 1968. The Migratory barren-ground caribou of Canada. Can. Wildlife Serv. Monograph, No. 3.

Kirk, G. S. and J. E. Raven. 1957. The Pre-Socratic Philosophers. Cambridge University Press.

Klein, David. 1972. "Toward an ecophilosophy." Tomte Symposium on Ecology and Land Use, Steinsgard, Norway.

Klein, David. 1983. Personal Communication.

Klein, David R. 1970. IUCN Publ. New Series No. 16:209-242.

Koch, M. 1996. Wildlife people and development." Trop. Ani. Health Prod. 28:68-80.

Koestler, A. and J.R. Smythies, Eds. 1969. Beyond Reductionism: New Perspectives in the Life Sciences. London: Hutchinson.

Koestler, A. 1978. Janus: A Summing Up. New York: Random House.

Kohr, Leopold. 1957. The Breakdown of Nations. New York: E. P. Dutton.

Kohr, Leopold. 1977. The Overdeveloped Nations: Diseconomics of Scale. New York: Schocken Books.

Koop, C. Everett. 1988. In the 1988 Report on Nutrition and Health.

Korten, David. 1995. 'The Tyranny of the Global Economy.' West Hartford, CT.
_____. 2006. The Great Turning. San Francisco: Berrett-Koehler.

Kozlovsky, Daniel G. 1974. An Ecological and Evolutionary Ethic New York: Prentice-Hall.

Krebs, Charles J. 1985. Ecology. 3rd ed. New York: Harper & Row.

Kropotkin, P.A. 1972. Mutual Aid: A Factor in Evolution. New York: New York University Press.

Krutch, J. 1970. The Best Nature Writing of Joseph Wood Krutch. New York: Pocket Books.

Kuhn, T. (1970) The Structure of Scientific Revolutions. Chicago: University of Chicago Press.

Kunstler, James H. 1994. The Geography of Nowhere. New York: Free Press.
_____. 2006. The Long Emergency. New York: Grove Press.

Lackner, S. 1984. Peaceable Nature. New York: Harper and Row.

Lamarck, Jean. 1963. Zoological Philosophy. trans. H. Elliot. New York: Hafner Publishing.

Lappe, Frances Moore and Joseph Collins. 1979. Food First: Beyond the Myth of Scarcity New York: Ballantine.

Laszlo, Ervin. 1972. Introduction to Systems Philosophy: Toward a New Paradigm of Contemporary Thought. New York: Harper Torch.

Laszlo, Ervin et al. 1977. Goals for Mankind. New York: E. P. Dutton.

Laszlo, Ervin. 1989. The Inner Limits of Mankind. London: Oneworld.

Laughlin, Robert B. 2011. Powering the Future. New York: Basic Books.

Lehmann, Scott 1981. "Do Wildernesses Have Rights?" Environmental Ethics 3:129-146.

Leopold, Aldo. 1945. "The Green Lagoons." American Forests. 51:414.

_____. 1949. A Sand County Almanac. And Sketches of Here and There. New York: Oxford University Press.

Levitt, Stephen D. and S.J. Dubner. 2006. Freakonomics. New York: William Morrow.

Lewin, K. 1951. Field Theory in Social Science. D. Cartwright, Ed. New York: Harper & Row.

Liebig, J. von. 1840. Chemistry in its Application to Agriculture and Physiology. London: Taylor and Walton. (cited in Odum 1970)

Lieth, Helmut F H. and Robert Whittaker, Eds. 1975. Primary Productivity of the Biosphere. New York: Springer-Verlag.

Likens, Gene E., Ed. 1989. Long-Term Studies in Ecology. New York: Springer-Verlag.

Lincicome, D.R. 1969. The Goodness of Parasitism: A New Hypothesis. Thomas C. Cheng, Ed. Aspects of the Biology of Symbiosis. Baltimore: University Park Press.

Loehle, C., 1989. Catastrophe theory in ecology: A critical review and an example of the butterfly catastrophe. Ecol. Modelling 49:125-144.

Lorenz, Konrad. 1952. King Solomon's Ring: New Light on Animal Ways. trans. M. K. Wilson. New York: Crowell.

_____. 1974. Civilized Man's Eight Deadly Sins. trans. M. K. Wilson. New York: Harcourt, Brace, Javonovich.

Lovejoy, A.O. 1964. The Great Chain of Being: A Study of the History of an Idea. Cambridge: Harvard University Press.

Lovejoy, Thomas and Richard Bierregaard. In Soule, Michael. 1986. Conservation Biology: The Science of Scarcity and Diversity. Sunderland: Sinauer Associates.

Lovejoy. Thomas E and D. C. Oren. 1981. "The minimum critical size of ecosystems," in W.D. Billings et al., Eds., Forest Island Dynamics in Man-dominated Landscapes. New York: Springer-Verlag.

Lovelock, J. E. 1979. Gaia: A New Look at Life on Earth. Oxford University Press, Oxford.

_____. Ages of Gaia. New York: Bantam Books.

_____. 1991. Healing Gaia: Practical Medicine for the Planet. New York: Harmony Books.

_____. 2007. Revenge of Gaia. New York: Basic Books.

_____. 2009. The Vanishing Face of Gaia. London: Allen Lane.

Lovins, Amory. 1977. Soft Energy Paths. New York: Harper & Row.

Lucas, Oliver. 1990. The Design of Forest Landscapes. New York: Oxford.

Lyle, J. T. 1985. Design for Human Ecosystems: Landscape, Land use and Natural Resources. New York: Van Nostrand Reinhold.

MacArthur, Robert H. 1972. Geographical Ecology: Patterns in the Distribution of Species. New York: Harper & Row.

MacArthur, R.H. and E.O. Wilson. 1967. The Theory of Island Biogeography. Princeton: Princeton University Press.

Maillat, Dennis, Bruno Lecoq, Florian Nemeti, Marc Pfister. 2009. "Technology District and Innovation: The Case of the Swiss Jura Arc." Informaworld.

Mandelbrot, B. B. 1982. The Fractal Geometry of Nature. New York: Freeman.

Mander, Jerry. 1991. In the Absence of the Sacred. San Francisco: Sierra Club Books.

Mander, Jerry, and Edward Goldsmith. 1996. The Case against the Global Economy: And Turn toward the Local. San Francisco: Sierra Club.

Mangold, Robert et al. 1993. Tree Planting in the United States—1993. Washington: USDA, Forest Service.

Margalef, Ramon. 1968. Perspectives in Ecological Theory. Chicago: University of Chicago Press.

Margulis, Lynn. 1974. Five kingdoms—classification and the origin and evolution of cells. Evol Biol 7:45-48.

Margulis, Lynn and Dorion Sagan. 1986. Microcosmos. New York: Summit Books.

Margulis, Lynn. 1991. Big trouble in biology: Physiological autopoiesis versus mechanistic neo-Darwinism. In John Brockman, Ed., Doing Science. New York: Prentice Hall.

Marsh, G. P. 1964. Man and Nature: Or Physical Geography as Modified by Human Action. Cambridge: Harvard University Press.

Maruyama, Magorah. 1978. Cultures of the Future. The Hague: Mouton.

_____. 1980. "Toward Cultural Symbiosis." In Evolution and Consciousness: Human Systems in Transition, pp. 198-213. E. Jantsch and C. H. Waddington, Eds. Reading, MA: Addison-Wesley Publishing Co.

Maser, Chris. 1994. Sustainable Forestry. Delray Beach: St. Lucie Press.

Maslow, A. H. 1968. Toward a Psychology of Being. 2nd ed. New York: D. Van Nostrand Co.

_____. H. 1971. The Farther Reaches of Human Nature. New York: Viking Press.

Mazrui, Ali Al Amin. 1976. A World Federation of Cultures: An African Perspective. New York: Free Press.

McArthur, Robert H. 1972. Geographical Ecology: Patterns in the Distribution of Species. New York: Harper & Row.

MacArthur, R.H. and E.O. Wilson. 1967. The Theory of Island Biogeography. Princeton: Princeton University Press.

McDonough, William and Michael Braungart. 2002. Cradle to Cradle: Remaking the Way We Make Things. New York: North Point Press.

McHarg, Ian. 1969. Design with Nature. Garden City: Natural History Press.

McKibben, Bill. 2008. Deep Economy: The Wealth of Communities and the Durable Future. New York: Holt Paperbacks.

McLuhan, Marshall. 1964. Understanding Media. New York: McGraw Hill.

Meadows, Donella. 2007. Thinking in Systems. Chelsea Green.

Meadows, Dennis. 1982. "Fallacies in resource planning," in Charles Hewett, T. Hamilton, and I. Anderson, Eds., Forests in Demand. Boston: Auburn.

Meeker, Joseph. 1974. The Comedy of Survival. New York: Charles Scribner's Sons.

Merchant, Carolyn. 1980. The Death of Nature: Women, Ecology, and the Scientific Revolution. San Francisco: Harper & Row.

Merleau-Ponty, Maurice. 1968. The Visible and the Invisible. Translated by A. Lingis. Evanston, Northwestern University Press.

Midgley, Mary. 1989. Wisdom Information & Wonder. New York: Routledge.

_____. 2002. Science & Poetry. London: Routledge.

Miller, George. 1956. The magic number seven plus or minus two. Psych. Rev. 6.

Miller, G. T. 1982.Living in the Environment. Belmont: Wadsworth.

Mollison, Bill. 1988. Permaculture: A Designers' Manual. Tyalgum, Australia: Tagari Publications.

Moran, Emilio F., Ed. The Ecosystem Approach in Anthropology. Ann Arbor: University of Michigan Press.

More, Thomas. 1982. Utopia. London: Penguin Books.

Morgan, Conwy Lloyd. 1925. Emergent Evolution. New York: Henry Holt and Co.

Mumford, Lewis. 1956. The Transformation of Man. New York: Harper and Row.

_____. 1961. The City in History: Its Origins and Transformations and its Prospects. New York: Harcourt, Brace and World.

_____. 1967. Technics and Human Development. San Diego: HBJ Book.

Myers, Nancy and Carolyn Raffensperger, Eds. 2006. Precautionary Tools for Reshaping Environmental Policy. Cambridge: MIT Press.

Myers, Norman. 1984. The Primary Source. New York: Norton.

Naess, Arne. 1972. The shallow and the deep, long-range ecology movement. A summary. Inquiry, 16: 95-100.

_____. 1987. Self-Realization: An Ecological Approach to Being in the World. The Trumpeter: Journal of Ecosophy, 4 (3), 35-42.

_____. 2008. The Ecology of Wisdom: Writings by Arne Naess. (Alan Drengson and Bill Devall, Eds.). Emeryville, CA: Counterpoint Press.

Naisbitt, John. 1984. Megatrends: Ten New Directions Transforming Our Lives. New York: Warner Books.

Ndubisi, Forster. 2002. Ecological Planning. Baltimore: Johns Hopkins.

Nicolis, G., and Prigogine, I. 1977. Self-organization in Non-equilibrium Structures. New York: Wiley.

_____. 1989. Exploring Complexity: An Introduction. New York: W. H. Freeman.

Nieman, H. et al, Transgenic farm animals, Rev.sci.Off.int.Epiz, 24:285-298.

Norberg-Hodge, Helena. 2009. Ancient Futures. San Francisco: Sierra Club.

Norberg-Schulz, C. 1971. Existence Space and Architecture. New York: Praeger.

Noss, Reed. And Allen Cooperider 1994. Saving Nature's legacy: Protecting and restoring Biodiversity. Washington, DC: Island Press.

Odum, Eugene P. 1970. "Optimum population and environment: A Georgian microcosm." Current History 58:355-366.

_____. 1971. Fundamentals of Ecology. 3rd Edition. Philadelphia: W.B. Saunders.

Odum, Eugene P., Clyde E. Connell, and Leslie B. Davenport. 1965. "Population energy flow of three primary consumer components of old-field ecosystems." Ecology 43(1):88-96.

Odum, Howard T. 1957. "Trophic structure and productivity of Silver Springs, Florida." Ecological Monographs 27(1).

Odum, Howard T. and Elisabeth C. Odum. 1981. Energy Basis for Man and Nature. New York: McGraw Hill.

Olson, Steve. 2002. Mapping Human History. Boston: Mariner Book.

O'Neill, R.V. et al., Eds. 1986. A Hierarchical Concept of Ecosystems. Princeton: PUP.

Ophuls, William. 1977. Ecology and the Politics of Scarcity. San Francisco: W. H. Freeman.

Organic salvation down on the farm. 2004. New Scientist, 184: p 9.

Ovington, J.D., Dale Heitkamp, and Donald Lawrence. 1963. "Plant biomass and productivity of prairies, savanna, oakwood, and maize field ecosystems in central Minnesota." Ecology 41(1):52-65.

Owens, Owen D. 1993. Living Waters. New Brunswick: Rutgers University Press.

Passmore, J. 1974. Man's Responsibility for Nature: Ecological Problems and Western Tradition. London: Duckworth.

Pepper, S. 1961. World Hypotheses. Berkeley: University of California Press.

Pedersen, D. 1996. Disease ecology at a crossroads: man-made environments, human rights and perpetual development utopias. Soc Sci Med, 43(5):745-58 1996 Sep.

Perry, David A. 1995. Forest Ecosystems. Baltimore: Johns Hopkins University Press.

Pielou, E. C. 1974. Population and Community Ecology: Principles and Methods. New York: Gordon and Breach.

Pimental, D., Houser J, Preiss E, et al. 1997. Water reserves: agriculture, the environment and society; an assessment of the status of water resources. 'Bioscience' 47: 97-106.

Pimm, Stuart L. 1991. The Balance of Nature. Chicago: University of Chicago Press.

Pirie, N. W. 1976. Food Resources. London: Pelican Books.

Polunin, Nicholas, Ed. 1980. Growth without Ecodisasters? New York: Wiley.

Portmann, Adolf. 1964. New Paths in Biology. New York: Harper & Row.

Prigogine, Ilya. 1977. Order Out of Chaos: Man's New Dialogue with Nature. New York: Bantam Books.

Prigogine, Ilya. 1980. From Being to Becoming. San Francisco: Freeman.

Public Health Service, Centers for Disease Control, 1975.

Rapoport, Amos. 1969. House Form and Culture. New York: Prentice-Hall.

Rapoport, Amos. 1982. The Meaning of the Built Environment. Beverly Hills; Sage Pubs.

Rappoport, Anatol. 1960. Fights, Games, and Debates. Ann Arbor: University of Michigan.

Rapport, D.J., C Thorpe, and HA Regier, 1979. Ecosystem medicine. Bul Ecol Soc Am 60:180-182.

Rapport, D.J., et al. 1985. "Ecosystem Behavior Under Stress," American Naturalist 125:617-640.

Rapport, D.J. 1995. "Ecosystem health: An emerging integrative science." In Rapport, DJ, CL Gaudet, and P. Calow, Eds. Evaluating and Monitoring the health of large scale ecosystems. pp. 5-34. Heidelberg: Springer.

Rees, Bill and Mathis Wackernagel. 1996. Our Ecological Footprint: Reducing Human Impact on the Earth. Gabriola Island: New Society Publishers.

Reich, Robert B. 2010. Afterhock. New York: Alfred A. Knopf.

Reichel-Dolmatoff, Gerardo. 1971. Amazonian Cosmos. Chicago: University of Chicago Press.

Reinheimer, H. 1910. Evolution by Co-operation: A Study in Bio-economics. NC: NP.

Relph, E. 1976. place and placelessness. London: Pion Limited.

Riewe, R. R. Changes in Eskimo utilization of arctic wildlife. In L. C. Bliss et al., Eds., Tundra ecosystems: A comparative analysis. Cambridge: University Press, 1981.

Rifkin, J. 1982. Algeny: The Last Magic.

Ritzer, George 1995. The McDonaldization of Society: An Investigation into the Changing Character of Contemporary Social Life. Thousand Oaks: Pine Forge Press.

Robbins J. 2001. 'The Food Revolution: How Your Diet Can Help Save Your Life and Our World', Newburyport, MA.

Roberts, Neil. 1989. The Holocene: An Environmental History. New York: Basil Blackwell.

Robinson J. 2004. 'Pasture Perfect: The Far-Reaching Benefits of Choosing Meat, Eggs, and Dairy Products from Grass-Fed Animals'. Vashon, WA, Vashon Island Press.

Rodin, L.E., N.I. Bazilevich, and N.N. Rozov. 1975. "Productivity of the world's main ecosystems." In: Productivity of World Ecosystems. Washington, D.C.: National Academy of Sciences.

Rodman, J. 1977a. "The Liberation of Nature?" Inquiry 20:83-145.

Rogers, Richard. 1997. Cities for a Small Planet. London: Faber & Faber.

Rolston, III, H. 1983. "Values Gone Wild." Inquiry 26:181-207.

Roszak, Theodore. 1972. Where the Wasteland Ends. Garden City, New York: Doubleday.

Roszak, Theodore, et al., Eds. 1995. Ecospychology. San Francisco: Sierra Club.

Sachs, Jeffrey D. 2008. Common Wealth. New York: Penguin Press.

Sage, B. 1981. Conservation of the tundra. In L. C. Bliss et al., Eds., Tundra ecosystems: A comparative analysis. Cambridge: University Press.

Sahlins, Marshall. 1968. Tribesmen. Englewood Cliffs: Prentice Hall.

_____. 1972. Stone Age Economics. Chicago: Aldine Publishing.

Sale, Kirkpatrick. 1980. Human Scale. New York: Coward, McCann and Geoghegan.

Salk, J. 1973. Survival of the Wisest. New York: Harper & Row.

Santos, Boaventura de Sousa. 1998. The Fall of the Angelus Novus: Beyond the Modern Game of Roots and Options, Current Sociology 46 (2) 81-118.

Savory, Alan. 1988. Holistic Resource Management. Washington: Island Press.

Schaller, G.B. 1972. The Serengeti Lion. Chicago: University of Chicago Press.

Schellnhuber, H.J. et al., Eds. 2004. Earth System Analysis for Sustainability. Cambridge: MIT Press.

Schlosser E. 2001. Fast Food Nation. Boston MA, Houghton Mifflin, 2001.

Schneider, Stephen H. 1997. Laboratory Earth. New York: Basic Books.

Schonewald-Cox, Christine M. 1983. Conclusions: Guidelines to management. In C. M. Schonewald-Cox et al., Eds., Genetics and Conservation. Menlo Park: Benjamin/ Cummings.

Schor, J.B. and B. Taylor, Eds. 2002. Sustainable Planet. Boston: Beacon Press.

Schrodinger, Erwin. 1946. What is Life? The Physical Aspect of the Living Cell. New York: Macmillan.

Schulze, E.-D. and H. A. Mooney, Eds. 1993. Biodiversity and Ecosystem Function. New York: Springer-Verlag.

Schumacher, E. F. 1973. Small is Beautiful. New York: Harper and Row.

Schweitzer, Albert. 1949. Out of My Life and Thought. New York: Henry Holt and Co.

_____. 1957. The Philosophy of Civilization. trans. C. T. Campion. New York: Macmillan Co.

Searles, H. 1962. The role of the nonhuman environment. Landscape (Winter 1961- 1962):31-34.

Shepard, Paul and D. McKinley, Eds. 1969. The Subversive Science. Boston: Houghton- Mifflin.

Shepard, P. 1973. The Tender Carnivore and the Sacred Game. New York: Scribner's Sons.

_____. 1974. Animal rights and human rites. The North American Review Winter.

_____. 1978. Thinking Animals. New York: Viking Press.

_____. 1982. Nature and Madness. San Francisco: Sierra Club.

Sierra Club. 1987. Survey. San Francisco: Sierra Club.

Simberloff, Daniel and L. G. Abele, 1976. "Island Biogeography Theory and Conservation Practice." Science 191 4224: 285-6.

Singer, Peter. 1981. The Expanding Circle: Ethics and Sociobiology. New York: Farrar, Strauss & Giroux.

Singer, S. Fred, Ed. 1971. Is There an Optimum Level of Population? New York: McGraw- Hill.

Skolimowski, Henryk. 1978. "Eco-philosophy versus the scientific world view." Ecologist Quarterly 3 (Autumn): 227-248.

_____. 1981. Ecophilosophy. Boston: Marion Boyars.

Slater, P. 1974. Earthwalk. New York: Bantam Books.

Smil, Vaclav. 2003. The Earth's Biosphere. Cambridge: MIT Press.

_____. 2008. Global Catastrophes and Trends. Cambridge: MIT Press.

Smith, Anthony D. 1991. The Ethnic Origins of Nations. Cambridge: Blackwell.

Smith J. M. 2006. 'Genetic Roulette: The Documented Health Risks of Genetically Engineered Foods', on website www.seedsofdeception.com.

Smith J.R. 1950. Tree Crops: A Permanent Agriculture. Washington: Island Press.

Smith, Maynard J. 1968. Mathematical Ideas in Biology. Cambridge: Cambridge University.

Smith, P.M., and Watson, R.A. 1979. "New Wilderness Boundaries." Environmental Ethics 1:61-64.

Smuts, Jan C. 1926. Holism and Evolution. Ann Arbor, MI: University Microfilms.

Snyder, Gary. 1969. Earth House Hold. New York: New Directions.

_____. 1983. The Coevolution Quarterly, Fall.

Sokoloff, J. 1985. The Politics of Food. San Francisco, CA: Sierra Books.

Soleri, P. 1969. Arcology: The City in the Image of Man. Cambridge: The MIT Press.

_____. et al. 2012. Lean Linear City: Arterial Arcology. Scottsdale: Cosanti.

Soule, Michael and Wilcox, B. A., Eds. 1980. Conservation Biology: An Evolutionary-Ecological Perspective. Sunderland, MA: Sinauer Associates.

Soule, Michael. 1986. Conservation Biology: The Science of Scarcity and Diversity. Sunderland: Sinauer Associates.

Spellerberg, I.F. 1991. Monitoring Ecological Change. Cambridge: University Press.

Speth, James G. 2008. The Bridge at the Edge of the World. New Haven: Yale.

Stanley, Steven. 1981. The New Evolutionary Timetable. New York: Basic Books.

Steiner, Frederick. The Productive and Erosive Palouse Environment. Pullman: Extension.

_____. 1991. The Living Landscape. New York: McGraw-Hill.

Stevens, Peter. 1974. Patterns in Nature. Boston: Little Brown.

Stone, Christopher D. 1975. Should Trees Have Standing? New York: Avon Books.

Stroganov, S.U. 1969. Carnivorous Mammals of Siberia. Springfield: US Dept. of Commerce.

Sturm, Andreas et al. The Winners and Losers in Global Competition: Why Eco-efficiency Reinforces Global Competitiveness: A Study of 44 Nations.

Tainter, Joseph A. 1990. The Collapse of Complex Societies. Cambridge: Cambridge University Press.

Tansley, A.G. 1935. The use and abuse of vegetational concepts and terms. Ecology 16:284- 307.

Tapscott, Don and A. D. Williams. 2010. Wikinomics. New York: Portfolio.

Thackera, John. 2005. In the Bubble: Designing in a complex world. Cambridge: MIT Press.

Thom, Rene. 1975. Structural Stability and Morphogenesis: An Outline of a General Theory of Models. trans. D. C. Fowler. Reading, MA: W. A. Benjamin.

Thomas, Keith. 1983. Man and the Natural World. New York: Pantheon.

Thomas, Lewis. 1975. Lives of a Cell. Toronto: Bantam Books.

Thompson, William I. 1974. Passages About Earth. New York: Harper & Row.

_____. 1976. Evil and World Order. New York: Harper & Row.

_____. 1981. The Time Falling Bodies Take to Light. New York: Harper & Row.

_____. 1987. Gaia: A New Way of Knowing. Political Implications of the New Biology. Great Barrington, MA: Lindisfarne Press.

Todd, N. J. and J. Todd. 1984. Bioshelters, Ocean Arks, City Farming: Ecology as the Basis of Design. San Francisco: Sierra Club Books.

_____. 1994. From Eco-Cities to Living Machines: Principles of Ecological Design. Berkeley: North Atlantic Books.

Todd, Nancy Jack. 2005. A Safe and Sustainable World. Washington: Island Press.

Toynbee, Arnold J. 1976. Mankind and Mother Earth: A Narrative History of the World. New York: Oxford University Press.

The Trumpeter. Various articles on ecosophy and ecoforestry. BC, Canada: Trumpeter.athabascau.ca.

Tuan, Yi-Fu. 1974. Topophilia. Englewood Cliffs: Prentice-Hall.

Uexkull, J. von. 1957. A Stroll Through the World of Animals and Men. Instinctive Behavior. C. Schiller, Ed. New York: International Universities Press Inc.

Ulanowicz, Robert E. 1986. Growth and Development: Ecosystems Phenomenology. New York: Springer-Verlag.

Ulanowicz, R. E. and W. M. Hemp. 1979. Toward canonical trophic aggregation. American Naturalist 114:871-883.

United Nations Development Programme. 1996. 'Urban Agriculture: Food, Jobs and Sustainable Cities', New York, UNDP.

USDA Soil Conservation Service. 1980. Soil Survey of Whitman County, Washington. Pullman: USDA.

Van der Ryn, Sim and Stuart Cowan. 1996. Ecological Design. Washington: Island Press.

Varela, Francisco. 1976. "Not one, not two." Coevolution Quarterly. Fall.

_____. 1979. Principles of Biological Autonomy. New York: North Holland.

Waddington, C.H. 1960. The Nature of Life. New York: Atheneum.

_____. 1971. The Evolution of an Evolutionist. Ithaca: Cornell.

Waldrop, M. Mitchell. 1992. Complexity. New York: Touchstone.

Wallace, Anthony F. C. 1966. Religion: An Anthropological View. New York: Random House

Waltner-Toews, D. and E. Wall. 1997. Emergent perplexity: in search of post-normal questions for community and agroecosystem health. Soc Sci Med, 45(11):1741-9 Dec.

Webb, Warren L., William K. Lauenroth, Stan R. Szarek, and Russell S. Kinerson. 1983. "Primary production and abiotic controls in forests, grasslands, and desert ecosystems in the United States." Ecology 64(1):134-151.

Weizsacker, Carl F. von. 1951. The History of Nature. London: Routledge & Kegan Paul.

Wells, Malcom. 1981. Gentle Architecture. New York: McGraw-Hill.

Westing, Arthur H. 1981. "A world in balance." Environmental Conservation 8(3):177-183.

Wheelwright, P. 1962. Metaphor and Reality. Bloomington: Indiana University Press.

Whittaker, R.H., F.H. Bormann, G.E. Likens, and T.G. Siccama. 1974. "The Hubbard Brook ecosystem study: Forest biomass and production." Ecological Monographs 44:233-252.

Whitehead, A.N. 1929. Process and Reality. New York: Macmillan.

_____. 1967. Science and the Modern World. New York: Free Press. P. 136.

_____. 1958. The Function of Reason. Boston: Beacon Press.

Willard, B. E. et al. 1977. Ethics of Biospheral Survival: A dialogue. In Growth Without Ecodisasters? pp. 505-535. N. Polunin, Ed. New York: John Wiley & Sons.

Williams C. 2000. The Environmental Threat to Human Intelligence a study funded by Britain's Economic and Social Research Council in its Global Environmental Change Programme, April 24.

Wilson, A.K, J.R. Latham, and R.A. Steinbrecher. 2006. 'Transformation-induced mutations in transgenic plants: Analysis and biosafety implications. Biotechnology and Genetic Engineering Reviews, 23, p 209-226.

Wilson, E.O. 1975. Sociobiology: The New Synthesis. Cambridge: Belknap Press.

Wittbecker, A. E. 1970. Eutopias: A Poetic Commonwealth of the Earth. Newark: Shamrock Press.

_____. 1976. The Poetic Archaeology of the Flesh. Wilmington: Mozart & Reason Wolfe, Ltd.

_____. 1976. "The psychology of catastrophe: Environmental deterioration and rapid social change." Proc. Marsh Inst. 1:1-17.

_____. 1983. An optimum human population based on NCP. Ecological Society of America annual meeting, Fargo.

_____. 1986. "The place of human society in wilderness." The Trumpeter. 3(3):34- 38.

_____. 1986. "Deep anthropology: Ecology and human order." Environmental Ethics 8(3):261-270.

Wittbecker, A. E. 1990. "Metaphysical implications from physics and ecology." Environmental Ethics 12(3):276-281.

_____. 1991. "An empowered United Nations: Proposals for cooperation and survival." Common Voice: 1(1):1-8.

_____. 1995. "Saving common places: The Palouse," Wild Earth 5(1):54-58.

_____. 1995. "Gigatrends in Forestry," International Journal of Ecoforestry 11(2/3):69-78.

_____. 1997. "Waldgedankenexperiment—Forest thought experiment," International Journal of Ecoforestry 12 (3/4):1-4.

_____. 1999. "Varieties of Interaction in nature," The Trumpeter, Spring (Web Edition: www.athabascau.ca/trumpeter).

_____. 2001. "Ecological Thought Experiments" Sofia Echo Vol. 5, Issue 31, Aug 3-9, p. 12 (and 8 others in a series, 2001-2002).

_____. 2002. REviewing REthinking REturning: Essays. Baltimore: Cambridge Books and www.ebooksonthe.net.

_____. 2003. Good Forestry From Good Theories & Good Practices: Essays. Baltimore: Cambridge Books and www.ebooksonthe.net

_____. 2007. Global Emergency Actions (For a Small Urban Industrial Planet). Sarasota: Urania Science Press.

Wolfe, L. M. 1945. Son of the Wilderness: The Life of John Muir. New York: Knopf.

Woodwell, George M. and Robert Whittaker. 1968. "Primary production in terrestrial ecosystems." American Zoologist 8:19-30.

Worldwatch Institute. Annual. State of the World. Washington, DC: Worldwatch

About the Author

Alan Wittbecker was educated at the universities of Virginia, Idaho, Washington, and Oregon, finishing courses in Astrophysics, Psychology. Geography, and English. He has degrees in Veterinary Medicine, Phenomenology and Ecosystem Ecology. He received his doctorate from International College, Los Angeles, where he studied with Michael W. Fox, John B. Cobb Jr, Arne Naess, Paolo Soleri, David Klein, Henryk Skolimowski, Paul Shepard, and Buckminster Fuller. He continued his postdoctorate education in landscape ecology, forestry, anthropology, conservation biology, wolf ecology, and zoology.

In 1991, Wittbecker founded SynGeo ArchiGraph, LLC, a firm specializing in global and regional designs; he created designs for several bioregions, as well as for international frameworks (pro bono). Two years later he set up the educational program for the Ecoforestry Institute, becoming an Instructor in 1994, journal Editor in 1995, and Director from 1997 to 2000. He has worked on public and private forests across North America. He previously worked as a Research Associate Ecologist for the G.P. March Institute from 1976 to 1992, in forest and laboratory research; he was Director (by rotation) for four separate years.

A veteran of the US Air Force, Wittbecker is also a returned Peace Corps Volunteer from Bulgaria, where he monitored wolves in the Central Balkan Mountains. He has used his education and interests to explore a spectrum of ecological applications, from research on forest pests to the political implications of the protection of species and habitats. When not engaged in preservation activities, he enjoys walking, swimming, reading, and drawing, at the River Farm Forest. You can reach him at:

aw@syngeo.org

*Figure. The author thinking
(Credit: M. DePasse).*

Colophon
Type: Garamond
Display Type: Gill Sans
Book & Cover Design: Rian Garcia Calusa Designs
Graphics: Alan Wittbecker (Unless credited)
Author Drawing: Merissa DePasse
Editing: J. Garcia B. of Rian Garcia Calusa
Hardware: Macintosh G5, HP 3310
Software: SimpleText, Adobe InDesign & Acrobat
Furious Charge & Entertainment: Pippi Frog
Spiritual & Material Support: Precious Woulfe

www.ingramcontent.com/pod-product-compliance
Lightning Source LLC
Chambersburg PA
CBHW051212170526
45166CB00005B/1857